5

ARMS AND
THE PHYSICIST

Masters of Modern Physics

Advisory Board

Dale Corson, Cornell University
Samuel Devons, Columbia University
Sidney Drell, Stanford Linear Accelerator Center
Herman Feshbach, Massachusetts Institute of Technology
Marvin Goldberger, University of California, Los Angeles
Wolfgang Panofsky, Stanford Linear Accelerator Center
William Press, Harvard University

Series Editor

Robert N. Ubell

Published Volumes

The Road from Los Alamos by Hans A. Bethe
The Charm of Physics by Sheldon L. Glashow
Citizen Scientist by Frank von Hippel
Visit to a Small Universe by Virginia Trimble
Nuclear Reactions: Science and Trans-Science by Alvin M. Weinberg
In the Shadow of the Bomb: Physics and Arms Control
  by Sydney D. Drell
The Eye of Heaven: Ptolemy, Copernicus, and Kepler
  by Owen Gingerich
Particles and Policy by Wolfgang K.H. Panofsky
At Home in the Universe by John A. Wheeler
Cosmic Enigmas by Joseph Silk
Nothing Is Too Wonderful to Be True by Philip Morrison
Arms and the Physicist by Herbert F. York

ARMS AND THE PHYSICIST

HERBERT F. YORK

The American Institute of Physics

AIP Press
American Institute of Physics
500 Sunnyside Boulevard
Woodbury, NY 11797-2999

Library of Congress Cataloging-in-Publication Data

York, Herbert F.
  Arms and the physicist / Herbert F. York.
    p.   cm.—(Masters of Modern Physics; v. 12)
    Includes index.
    ISBN 1-56396-099-0
    1.   York, Herbert F.   2.   Arms race—United States—History.
    3. National security—United States—History.   4.   Physicists—United States—
  Biography.   I.   Title.   II.   Series.
    QC16.Y67A3   1994   94–13099
    327.174—dc20   CIP

This book is volume twelve of the Masters of Modern Physics series.

Contents

About the Series

Masters of Modern Physics introduces the work and thought of some of the most celebrated physicists of our day. These collected essays offer a panoramic tour of the way science works, how it affects our lives, and what it means to those who practice it. Authors report from the horizons of modern research, provide engaging sketches of friends and colleagues, and reflect on the social, economic, and political consequences of the scientific and technical enterprise.

Authors have been selected for their contributions to science and for their keen ability to communicate to the general reader—often with wit, frequently in fine literary style. All have been honored by their peers and most have been prominent in shaping debates in science, technology, and public policy. Some have achieved distinction in social and cultural spheres outside the laboratory.

Many essays are drawn from popular and scientific magazines, newspapers, and journals. Still others—written for the series or drawn from notes for other occasions—appear for the first time. Authors have provided introductions and, where appropriate, annotations. Once selected for inclusion, the essays are carefully edited and updated so that each volume emerges as a finely shaped work.

Masters of Modern Physics is edited by Robert N. Ubell and overseen by an advisory panel of distinguished physicists. Sponsored by the American Institute of Physics, a consortium of major physics societies, the series serves as an authoritative survey of the people and ideas that have shaped twentieth-century science and society.

Preface

Agreat technological arms race began with the discovery of nuclear fission in 1938 and ended with the collapse of the Soviet Union in 1990. The essays in this volume present my historical and political analyses of most of the major events that made up that arms race. I joined the arms race soon after it started (I was recruited into the Manhattan Project directly from college in November 1942) and I was still actively engaged in it when it suddenly ended almost fifty years later. These analyses are, therefore, not those of a dispassionate scholar. Rather, they are "my version of the facts," as Leo Szilard would have put it, as I experienced them in real time during a half century of close observation of, or direct participation in, the decision-making process that guided the American side of the race.

But my views of these events were not formed solely out of my personal experiences with them. They were, like the views of everyone else, also strongly informed by all the prejudices (I do not regard this term as having wholly negative connotations) I picked up as I went along, including some formed well before the arms race began.

Some of these prejudices rose out of large-scale events that took place during my formative years. Many of these can be characterized just by giving the names of certain places: Munich, Pearl Harbor, Auschwitz, Yalta, Hiroshima. Others, at least as important, are the result of what I learned from other people. At first, these were people I never knew or ever hoped to meet; later they were teachers, colleagues and friends, or more generally, but sometimes loosely, mentors of various kinds. Given the times in which I lived and worked, it will be no surprise that all the important ones were men.

The first person outside of my family who influenced my thinking was Franklin D. Roosevelt. I recall the excitement—and suspense—inherent in all four of his elections, and I participated in the last one. I was an avid newspaper reader from as early as I can remember, and so I was, even as a young teenager, very much aware of the revolution he was working in the way government related to the People generally, and the Economy in particular. I wouldn't have put it in those particular words at the time, but it was with great interest that I watched, in person, the parades of working men supporting his ideas and hopes, and observed through *Movietone* newsreels many of the other big events that took place as he developed his programs for coping with the Great Depression and for putting people "back to work" (my father and many other relatives were either unemployed or found work only on government-sponsored programs for many months, sometimes years, in those days). Later, of course, I followed and admired his career as the leader of America at war.

I also learned in real time from Adolph Hitler, Benito Mussolini and Jozef Stalin how terrible things can be and just how thin the veneer of civilization that protects us really is. I clearly recall hearing on the radio when I was eleven that von Hindenburg, then newly elected president of Germany, was going to appoint Hitler to be Chancellor. Images of Hitler standing before great crowds with his arm stuck out are part of my personal memory—not just a part of "History"—and so are the responses of the crowd: "Sieg Heil! Sieg Heil." It was the same with Mussolini, except that in his case his chin stuck way out, too, and the crowd's answer was "Duce! Duce!" As a kind of counterpoint, I watched Haile Selassie standing alone before the League of Nations pleading for help in containing Mussolini's invasion of his helpless country. With Stalin it was similar. The newsreels often showed him standing on top of that famous Lenin mausoleum with an enigmatic smile on his lips as enormous squads of soldiers and flag bearers marched by. Although later we and the Soviet Union became allies in the war against fascism, I never completely put aside the certain knowledge that Stalin had killed and imprisoned millions of innocents in order to "build Communism." And finally, during the last of my high school years, there was the Spanish Civil War that brought all this together and should have warned us of what was yet to come.

Some of the books I read as a teenager also penetrated deeply into my psyche. Many were biographies, the most memorable being about Galileo, Columbus, Louis Pasteur, Thomas Edison, and Alexander Graham Bell. Perhaps the most important was *Astronomy for Amateurs* by the French astronomer Camille Flammarion, a book that was given to me by a slightly eccentric uncle sometime before I was ten years old. Originally entitled *Astronomie des Dames,* and based largely on 19th century ideas, it was purely descriptive and thus possible for me to understand at that

age. It was written in very flowery language, especially suitable, I suppose, for *les dames* and the small boys of those days, and it was replete
with the captivating (for me) large numbers that are necessary to present
the sizes, distances, and time scales involved. I read the book through
more than once. I learned many of those wondrous numbers by heart and
I still remember many of them.

The earliest of my important personal mentors was Victor Weisskopf. I
met him in the fall of 1941, only a few months before Pearl Harbor. He
had arrived in this country a few years earlier as a refugee from the Hitler
Terror and was still a relatively new and young member of the physics
faculty at the University of Rochester. I was a third-year undergraduate,
just then switching my major to physics. As all physicists are now aware,
he was a very warm and thoughtful person. He was always ready to be
helpful to students, and his genuine love for science showed through
clearly and inspired the same in others. Perhaps most important, I knew
even then that he had personally worked with the persons who had made
the discoveries, elaborated the theories, and written the first articles and
books about the great physics I was then engaged in learning. He was, in
brief, the first "world scale" person I ever met and I literally enjoyed the
privilege of knowing him. Thus, having been so early sensitized to his humanity and wisdom, it was easy for me in later life to be strongly influenced by his views about the arms race and his belief in the need to deal directly with the Soviets in seeking a way out of it the dangers inherent in it.

By 1942/43 the world situation had developed to a point such that everyone had to drop whatever else they might be doing and contribute his
individual effort to winning the war as soon as possible. I briefly considered working on several different war research projects, but I finally decided on the one at the University of California in Berkeley for two principle reasons. One was that Ernest O. Lawrence headed the project there
and he had been a remote hero to me ever since I first learned several
years earlier about his invention of the cyclotron. The other was that California was home to the Golden Gate, and to the biggest trees and the highest mountains in the forty-eight states, and I wanted to see them all as
soon as I could (I had never before been west of Buffalo, N.Y.).
Lawrence—and California—fully lived up to my expectations and he
soon became not only a close-up hero, but also a sort of father figure (he
was exactly 20 years older than I). Over the course of the next fifteen
years, I found—or made—many occasions to talk with him about what I
was doing at the moment. Although I learned most of my physics from
others—Oppenheimer and Segre to name two—I learned how to organize
and lead "big physics" programs from Lawrence himself. He evidently

liked what he saw in me, and in 1952 he assigned to me the job of establishing what the world now knows as the Lawrence Livermore National Laboratory. During my tenure as director at LLNL (1952-58), he almost never told me what to do but even so—or perhaps for that reason—I always strived to run the place as closely as I could to the way he ran the Berkeley lab.

The Livermore Lab had been founded, in part, because of Edward Teller's strong dissatisfaction with the way things were going at the older Los Alamos Lab and his consequent—and highly personal—political campaign for the establishment of a second, competitive lab somewhere else. As a result, he joined Lawrence's new enterprise at the very start and he has used it as his main base of operations ever since. (One of the later essays in this book describes in detail how all this came about.) For the next five years, I worked very closely with Teller and came to understand in detail his world view. At first, I accepted most of it, but eventually his extreme views about the need for ever more and better nuclear weapons and his total hostility to any kind of negotiations with the Soviets, helped me to realize (even before I left the laboratory) that, on the contrary, there simply had to be other, additional, approaches to our (very real) security problems, and that negotiations were in fact one of the possible means for dealing with them.

The American approach to managing all government-sponsored R&D, especially that part of it related to national defense, was largely reorganized after the Soviets launched the first Sputnik in October 1957. One serendipitous result was that I found myself in Washington for the next four years occupying in succession several high-level positions responsible for making policy in all areas of defense and space R&D. These responsibilities brought me into direct and frequent contact with President Eisenhower and his last two science advisors, James Killian and George Kistiakowski. My frequent meetings with these three men, especially Eisenhower, brought about the final step in the evolution of my views on national security and the role of nuclear technology in supporting it. Specifically, I came to understand that the nation—and the world—faced two serious problems simultaneously. The popular name for one of these was "the Soviet Threat" and the popular name for the other was the "nuclear arms race." They both had to be faced and dealt with, but the easy solutions to each one of them only made the other worse. I came to understand that, indeed, there probably were no simultaneous complete solutions to both of them, and that, therefore, the basic task of the leaders was simply to manage them as well as they could, all the while maintaining both international peace and national security until the world situation could evolve into a form such that these problems could be solved together. (One of the later essays in this book deals specifically with conversations I had with Eisenhower on these issues after he retired.)

Among my contemporaries, Harold Brown has influenced my life and my thinking the most. We first met when he came to Lawrence's laboratory directly after earning his degree at Columbia University in 1949. As luck would have it, he joined me immediately in some work I was doing related to the accelerator production of plutonium and tritium and then, a little later, in preparing a diagnostic experiment to be carried out in the Pacific during the first ever thermonuclear explosion. Three years after that, in 1952, he was one of my principal assistants in establishing the Livermore laboratory and working out its program for the next five years. In 1961, at the end of my first tour of duty in the Pentagon, I recommended to Secretary of Defense McNamara that Harold be appointed as my successor. He took my advice, and the rest, as they say, is History. Throughout the years since, I have kept in close touch with Harold and I still make it a regular practice to check out my views on national security with him as often as I can, especially when something dramatic or decisive is happening (and there has been a lot lately!).

OVERVIEW

Making Weapons, Talking Peace

My earliest memories of public affairs include the election of Herbert Hoover, the Great Depression, the election of Franklin Roosevelt, the rise of Adolf Hitler, the Japanese invasion of China, and the Italian invasion of Ethiopia—all of which took place while I was still in grade school. They continue with the Spanish civil war, which occurred when I was in high school. The European branch of World War II started in the same month that I entered college at the University of Rochester.

Soon after that Pearl Harbor came along, and my professors began to leave in order to contribute to the war effort. I eventually joined them by going off to Berkeley and the Manhattan Project.

When the war ended, I was demobilized, so to speak, along with tens of millions of others all over the world, and I began what I hoped and thought was going to be a normal peaceful career in pure science. It was not to be. After only three and a half years, major external events, including the explosion of the first Soviet atomic bomb and the Korean War, brought me back into the nuclear arms race, and my life has been largely caught up in it ever since. This is a chronicle of my involvement, beginning with my work on the Hiroshima bomb and ending with my service as Jimmy Carter's chief negotiator at the Comprehensive Test Ban talks in Geneva.

It begins in November 1942, just before my 21st birthday, when a recruiter from the Berkeley Radiation Laboratory found me at the University of Rochester, where I was a first-year graduate student in the department of physics. Sidney Barnes, a Rochester physics professor who had gone to Berkeley in early 1942, had given my name to a recruiter, and so when he came our way later that year, he was looking for me specifically.

I could not have remained a graduate student any longer even if I had wanted to. Like most of my contemporaries, I wanted to do my part to win the war. I flirted briefly with other war projects—at MIT, Columbia, and McGill—but the lure of Berkeley was by far the strongest. I had recently started doing research using a special machine called a cyclotron, which Ernest O. Lawrence, the director of the Berkeley laboratory, had invented, and the idea of one day going there and joining him had been forming in the back of my mind for some time.

Lawrence's Berkeley Radiation Laboratory was the only one of the major institutions making up the Manhattan Project that was actually founded by its director. All the other units were created at the time by official fiat, and all the other directors were appointed by higher authority. In marked contrast, Lawrence's lab was already more than a decade old when it joined the project to make the bomb. This unique situation gave Lawrence an extra measure of leverage in his dealings with higher government and university authorities both then and after the war.

Like their colleagues everywhere, the physicists at Berkeley immediately realized that the discovery of fission in 1938 was of transcendent importance. Lawrence's younger coworkers quickly confirmed the new discoveries and expanded on them. Among other things, they discovered and determined the properties of plutonium, one of the two materials suitable for the making of nuclear bombs.

Lawrence followed these events very closely, but he devoted his personal energies to three other activities. One of these was the promotion of his plans for still bigger cyclotrons. Another was to follow developments in the expanding studies of uranium fission and to make recommendations about what should be done in the area. The third was the conversion of his laboratory from one engaged in peacetime research to one dedicated entirely to the uranium project. He pushed hard to focus attention on the possibility of building a bomb, rather than just on that of generating power for ship propulsion or similar purposes; he argued strongly that the production of both U-235 and plutonium should be pursued in parallel; and he very early on urged that the project be greatly expanded to bring in the large number of scientists and to build the institutions necessary to accomplish the task quickly.

Lawrence got the responsibility for "small-sample preparation, electromagnetic separation methods and certain experimentation [on plutonium]." In anticipation of this assignment Lawrence had already disassembled one of his smaller cyclotrons in order to use its magnet as the basic building block for a device capable of producing microgram quantities of U-235. In addition, he diverted the much larger magnet originally intended for his next and biggest cyclotron to the construction of a still larger device, later dubbed the Calutron, for producing multi-gram quantities of the same material.

In the summer of 1942 the great 184-inch magnet was turned on and experimental prototypes of the Calutron, Lawrence's gigantic mass spectrometer, were placed in operation between its pole pieces. By that time much of Lawrence's senior staff had left the laboratory to help launch other war-related projects. They were always sent off with his blessing and, usually, at his urging. As a result, a largely new and mostly even younger group had to be recruited to carry out the work at Berkeley.

Accordingly, my first assignment was as a member of the crew of the R-1 Calutron, one of two prototypes sitting between the pole pieces of the 184-inch magnet. The cochiefs were Frank Oppenheimer (Robert Oppenheimer's younger brother) and Fred Schmidt. Our task was to study the operating characteristics of the R-1 and, by using the Edisonian cut-and-try approach, to maximize its ability to extract U-235 in as pure a form as possible from natural uranium.

The details of all this came as a complete surprise to me, even though I had been quite certain even before my arrival in Berkeley that the Radiation Laboratory was engaged in some way in making a nuclear bomb. Nobody had told me that in so may words, but the general idea of nuclear energy and nuclear bombs had been in the air in the early 1940s. Given Lawrence's reputation in nuclear physics, it was easy to put two and two together and conclude that Berkeley was somehow involved in making uranium bombs.

As clearly as I can recall, the word *uranium* was breathed in my ear only once after my arrival. From then on, code words were used for everything of special relevance to the project. Uranium itself was called tuballoy, a code name I later learned had been invented by no less than Winston Churchill. The three natural isotopes of uranium were called W (U-224), X (U-235) and Y (U-238). Mixtures of isotopes especially enriched in U-235 were called R, and the residue of material depleted in U-235 was called Q. The object of the whole enterprise was to produce R as fast as possible and to have it be as rich in X as we could make it.

At one point I told Lawrence I could to a better job if I could know the trade-off between quality and quantity of uranium. After his next trip to Los Alamos, he handed me a small slip of paper on which was written the critical mass of enriched uranium (in arbitrary units) for four or five different levels of U-235 concentration (all, of course, in numbers only, without any of those forbidden words). Because of the compartmentalization of information, I was the only junior person in our part of the project who had this information. I did make use of it in designing product collectors in order to optimize the rate at which the plant approached the production of critical mass.

After the war ended, I convinced myself that my efforts, while not at all profound or even particularly clever, had had a small effect on the outcome. I had helped either to make the Hiroshima bomb available a few days earlier or to make its explosion a little more powerful. I knew, of course, that hundreds of others could make similar statements about hundreds of other small contributions, but I was pleased with what I had been able to do.

I do not recall anyone preaching to me or others about the need for secrecy. The need was explained just once—but firmly. Thereafter the whole atmosphere of the time and place strongly reinforced it. To my recollection, following my first day I never again heard the word *uranium* either in a normal conversation or in a confidential aside. This custom became deeply ingrained in me and everyone I knew. As a result, after news of the bomb burst upon the public two and a half years later, it was deeply shocking for me to read that forbidden word in the headlines and to hear people utter it out loud—with a certain awe, to be sure, but nonetheless as if it were just another, normal word. Hearing about the bomb was one of those things that caused a sudden, queasy feeling in the pit of my stomach.

In recent years, in classes and special lectures, I've had many occasions to describe to younger people the project, the bomb and its use. I've found that at the start a very wide gap separates us. The first thing most of my listeners learn about World War II is that we won it. That is, so to speak, the last thing I learned about it. The first thing they learn about the atomic bomb is that we dropped one on Hiroshima and another on Nagasaki. This is the last thing I learned about the project. For most people born after 1940, those events marked the beginning of the nuclear arms race with the Soviets. For those of us in the project, they heralded the end of history's bloodiest war.

Two especially important US authorities—Vannevar Bush and General Leslie Groves, who had directed the Manhattan Project—estimated that achieving the bomb would take the "backward" Soviets much longer than it had taken us—decades at the very least. Nearly everyone else—veterans of the US project, professional intelligence analysts, even Winston Churchill—estimated that four years would be enough for the Soviets to end the American monopoly. In 1947 the intelligence estimates said the first Soviet atomic bomb would come in another two years—that is, in 1949—but at that point the estimating process got stuck in a rut. As more time passed, the estimate continued to be "two more years" rather than "in 1949." Thus, on the eve of the actual event the latest estimate was still "two more years." In the context of this static prediction, the Soviet explosion was, literally speaking, a surprise.

It was also an exceedingly ominous event, one widely seen as bringing serious new dangers of a kind totally different from any we had ever faced before. The Soviet atomic bomb, combined with the projected acquisition of very long-range aircraft, promised the end forever of our historical invulnerability. Almost immediately, serious concern over the possibility of a devastating surprise attack on us rose within nuclear and military—especially Air Force—circles.

Lawrence's reaction to the Soviet atomic bomb was easily predictable. Seven years before, he had led the call for an active response to the possibility of a German atomic bomb, and now he reacted similarly to the reality of the Soviet bomb. A new nuclear threat seemed to call for a nuclear response, and so he actively sought ways to reinvolve his laboratory in its development. Shortly after the explosion in Central Asia, Lawrence and Luis Alvarez set out on a trip to Washington to explore the question of what the Rad Lab might do. They traveled east by way of Los Alamos in order first to visit with the laboratory director, Norris Bradbury, as well as with Edward Teller, then a visitor to Los Alamos on leave from the University of Chicago, and others who, they believed, had to be similarly concerned about the new turn of events.

Teller told them that the hydrogen bomb was the proper answer to the new challenge. This type of bomb, often also called the superbomb because its power was estimated to be perhaps a thousand times that of an "ordinary" fission bomb, had preoccupied Teller since the earliest days of the Los Alamos project. He had wanted to push ahead with it even then, but he was frustrated in this desire by Robert Oppenheimer's insistence that all efforts be concentrated on the fission bomb. Oppenheimer believed that a fission bomb was much more likely to be produced in time to be of use during the war. Besides, it would be needed as the trigger for a hydrogen bomb in any event. Teller had only grudgingly accepted Oppenheimer's conclusions during the war, but ever since its end he had been trying to find a way to get serious work going on his pet project. The challenge of the Soviet bomb seemed to provide the impetus that was previously lacking, and he used every means and argument he could think of to exploit it.

At the time it was still not known exactly how a hydrogen bomb might be constructed, only that in principle a few barrels of liquid deuterium, perhaps laced with tritium, would produce a prodigious amount of explosive energy if it could be heated to a high enough temperature for a long enough time—that is, to a temperature of many tens of millions of degrees for a fraction of a microsecond, quite a long time for such extreme temperatures and pressures.

Teller had never been satisfied with the management of the American nuclear weapons program or felt that the total effort devoted to it was adequate. His conflict with Oppenheimer during the war was notorious, as was his refusal to participate in the postwar continuation of the work at Los Alamos unless the government and the laboratory would both commit themselves to a substantial nuclear test program. In the fall of 1949, Teller had fought the General Advisory Committee (GAC) of the Atomic Energy Commission over the issue of whether the US should initiate a high-priority program to develop and build superbombs. He was even disappointed when, on 31 January 1950, President Truman ordered the AEC to "continue its efforts on...the so-called hydrogen or superbomb." Truman's use of the word *continue* rather than *charge ahead* had led Teller to feel—quite mistakenly—that his cause had been lost. Even after the Los Alamos laboratory had accelerated and expanded its program in response to the President's directives, Teller continued to believe that laboratory leaders—Bradbury and Carson Mark, among others—were not doing enough to recruit new people and were not diverting a sufficient fraction of the laboratory's effort from fission bomb development to the search for the super.

In mid-1951 all this strife finally came to a head. That May the George shot had proved that a mixture of deuterium and tritium really would explode if the right initial conditions could be achieved. It showed that a superbomb could be built, and Los Alamos set out reorganizing itself to do so in the shortest possible time. Bradbury, surely with Washington's concurrence, decided to put longtime senior laboratory staffers, and not Teller, in charge of the program. In Bradbury's judgment, Teller was no manager, and now that the basic idea was in hand, people who could manage and orchestrate a complex effort involving many different technologies had to be put in charge. More important, Teller's strident complaints about the laboratory management had so poisoned the atmosphere that it was simply impossible to give him the kind of autonomous position within the laboratory that he was demanding.

For Teller this was the last straw. Soon after these new organizational arrangements were announced, he left Los Alamos and returned to the University of Chicago. His memoir on the subject made it clear that one of his several reasons for leaving was his desire to press his campaign for a second laboratory.

Teller took his message to friends on the Congressional Joint Committee on Atomic Energy. There he found an especially sympathetic listener not only in the chairman, Senator Brian McMahon, but also in the chief staff aide, William L. Borden—the latter being the man who three years later formally accused Oppenheimer of being a Soviet agent.

While Teller was lobbying in the defense establishment and in Congress, he was also doing what he could within the Washington headquarters of the AEC itself. He found some support for his ideas from Gordon Dean, who had recently succeeded David Lilienthal as chairman of the AEC, and from Thomas Murray, one of the newer commissioners. Following the usual procedures in such matters, they referred the idea of a second laboratory to the GAC for consideration at its December 1950 meeting and again at several later meetings. Except for Willard F. Libby, a University of Chicago chemistry professor and personal friend of Teller's, the GAC members took a dim view of the idea of a new laboratory and recommended against establishing one. They were influenced in part by Bradbury, who said it wasn't needed and would inevitably divert essential support away from Los Alamos. In addition, many of the members of the GAC had previously found themselves in protracted conflicts with Teller over related issues, and that probably influenced their attitudes toward his latest proposal. A year later, in December 1951, after the GAC had made its third or fourth negative recommendation on the matter to the AEC commissioners, Murray put in the call to Lawrence that in turn prompted a New Year's Day query to me about my views on the matter.

I did not have enough information at hand to reach a firm conclusion, so I went off on a series of trips to sound out the situation. I visited Los Alamos to learn the views of my friends and colleagues there, and then I traveled on to Chicago to have a thorough discussion with Teller. The responses came out as one would suppose: The Los Alamos people felt they could do all that was required, and Teller was convinced they could not. I went on to Washington, where I met with the same AEC and Air Force officials Teller had seen. Everyone in the Air Force seemed to think that not enough was currently being done to exploit the recent breakthrough and new institutional arrangements were needed to set things right.

I also visited John Wheeler at Princeton University. He had long been involved in nuclear weapons and the search for the super, and he had recently set up a small group of theoretical physicists based in the nearby Forrestal Center, to help Los Alamos carry out some of the key calculations needed to convert the invention by Teller and Stan Ulam from a set of sketches and theories into a workable weapons system. I found Wheeler very much in agreement with Teller.

My early reports to Lawrence on the results of my travels confirmed his own preliminary conclusions that a second laboratory was needed. Combining his own prejudgments with what Teller and I had told him,

Lawrence informed his friends at the AEC that he would, if they wished, take on the task of establishing an additional weapons research center at Livermore, California, as a branch of the Radiation Laboratory and that he could staff it, at least at first, largely with people already on the payroll. This proposal changed the situation radically. It clearly meant much less initial expense and an immediate, if small, cadre of people ready to go to work as needed. As Oppenheimer later recalled, the GAC and the AEC very quickly "approved the second laboratory as now conceived because there was an existing installation, and it could be done gradually and without harm to Los Alamos."

Lawrence asked me to draw up plans for the new research center. In response I began to sketch out my ideas of how to go about it. After a few weeks Lawrence asked me if I thought I could run it. After only an overnight hesitation I told him it was worth a try, and he simply instructed me to do so. It really was that casual—no search committee or any of the other procedures to which we are now accustomed. In keeping with his standard style of operation, Lawrence gave me no new title, no immediate raise in salary or any other change in status.

Finally, and in close accord with Lawrence's views of the matter, the AEC in June 1952 approved the establishment of a branch of the Berkeley laboratory at Livermore to assist in the thermonuclear weapons program by conducting diagnostic experiments during weapons tests and performing other, related research. The question of how soon, or even whether, the Livermore Laboratory would actually engage directly in weapons development was left open, however. Teller was extremely dissatisfied with the vagueness of the AEC's plans for the new laboratory.

Finally, after mulling the matter over, Teller, in the course of a well-lubricated reception held at the Claremont Hotel in Berkeley in early July to celebrate the launching of the new enterprise, suddenly announced to Lawrence, Gordon Dean and me that he would have nothing further to do with the plans for establishing a laboratory at Livermore. Lawrence was prepared to go ahead anyway, and he even suggested privately to me that we would probably be better off without Teller. However, at the insistence of Captain John T. Hayward (then deputy director of the AEC's Division of Military Applications), intense negotiations were resumed among all concerned. Within days this led to a firm commitment by Dean that thermonuclear weapons development would be included in the Livermore program from the outset, as well as to a renewed commitment on the part of Teller to join the laboratory.

On 2 November 1952, the world's first large thermonuclear device, codenamed Mike, exploded on Elugelab Island at Eniwetok Atoll. Its yield was 10.5 megatons, fulfilling the prediction that the superbomb, if it could be made, would be a thousand times more powerful than the bomb dropped on Hiroshima. On the basis of ideas by Teller, Ulam and others, it was built and tested by the Los Alamos Laboratory. Because his relations with the Los Alamos leaders were so severely strained, Teller did not accompany the lab's team to observe it.

AEC authorities, on instructions from the White House, clamped an extraordinarily tight curtain of security over the whole operation. They intended to allow no post-test reports to be sent from the Pacific back to the laboratories or anywhere else until after there had been an on-the-spot analysis of what had happened. Even then, the first word would go directly to Washington only. The Task Force Command did, however, broadcast a coded signal that indicated the moment when the button had been pushed. Because of some experiments on long-range effects that we were doing at Livermore, we were given the means for decoding that message. The moment I received it at my office at the lab, I noted the time and telephoned Teller, then standing by at the Berkeley seismometer, to tell him just when "zero hour" had passed. He kept a very close watch on the seismometer, and at the appropriate time, some 14 minutes after zero hour, he saw the needle jump. He called me to say, "It's a boy!"

When I reflect back on that moment, as I sometimes do in preparing or giving lectures on the history of the nuclear era, a feeling of awe and foreboding always recurs. Even at the time, I thought of that moment and of that coded message as marking a real change in history—a moment when the course of the world suddenly shifted, from the path it had been on to a more dangerous one. Fission bombs, destructive as they might be, were thought of as limited in power. Now, it seemed, we had learned how to brush even those limits aside and to build bombs whose power was boundless.

Could the development of the H-bomb have been avoided? I do not see how. Even if there had been no other reasons—and there were others— the almost total lack of communications between Stalin's Russia and the rest of the world would alone have made its avoidance impossible as a practical political matter.

The birth of Mike led inevitably and immediately to a demand for developing practical versions of the new bomb. It settled, once and for all, Livermore's role in the program. There was obviously plenty of work ahead for both labs.

Our working philosophy, which I set out at the very beginning and which everyone readily accepted, called for always pushing at the technological extremes. We did not wait for higher government or military

authorities to tell us what they wanted and only then seek to supply it. Instead, we set out from the start to construct nuclear explosive devices that
had the smallest diameter, the lightest weight, the least investment in rare
materials, the highest yield-to-weight ratio, or that otherwise carried the
state of the art beyond the currently explored frontiers. We were completely confident that the military would find a use for our product after
we proved it, and that did indeed usually turn out to be true.

In keeping with this philosophy, I at one point proposed to the AEC
that we build and explode a bomb considerably bigger—more than 20
megatons—than any built before. According to procedures then in effect,
the President personally approved or disapproved every test. In the case
of this one, I was informed that when it was presented to President Eisenhower he said, "Absolutely not. They are already too big." Years later,
Andrew Goodpaster, who had been the President's military assistant, told
me it was during this period that Eisenhower concluded, "The whole thing
is crazy. Something simply has to be done about it."

Eisenhower had long been concerned about the nuclear arms race. He
was convinced it was leading us somewhere we had best not go, and he
looked for ways of getting it under control. The proposals presented in his
"Atoms for Peace" speech had been an early stab at the general issue.
Later he proposed "open skies" over Russia and America as a means for
reassuring each country that no grand surprises were being prepared deep
inside the other's territory. And ever since the "Bravo" test in the Pacific
in 1954, when 23 Japanese fishermen suffered radioactive fallout on
board their tuna trawler, Daigo Fukuryu Maru (Fortunate Dragon Number
5), Eisenhower had mulled over the possibility of a nuclear test ban both
as a solution to the fallout problem and as a means for slowing down the
arms race. Adlai Stevenson had raised the issue of fallout during the 1956
Presidential campaign in a way that Eisenhower took as a personal attack,
and he backed off from his own tentative thrusts in order not to be seen as
acquiescing to his opponent's charges. By mid-1957, however, he was
again considering the possible merits of a nuclear test ban as a "first step"
on the way to a more comprehensive arrangement. Premier Khrushchev
apparently was thinking along similar lines independently. Thus, even before the furor over Sputnik had faded, Eisenhower turned to his new
President's Science Advisory Committee to give him advice.

During my period of full-time service in the Pentagon, first as the chief
scientist of the Advanced Research Projects Agency and later as the director of defense research and engineering, I became more and more committed to the view that the nuclear arms race must somehow be brought
under control and that a nuclear test ban would make a good first step.

The general climate on the west side of the Potomac ranged from dubious to hostile. As a result, I found myself in the unusual and faintly amusing situation of being a member of a rather special minority in the national security establishment, one that included the commander-in-chief himself.

This situation had its dark side as well. On a memorable occasion in early 1960, a meeting took place between Secretary of Defense Thomas Gates, Chairman John McCone of the AEC, myself and one or two others. At one point the discussion turned to the question of whether the Soviets were cheating on the nuclear test moratorium, then still in effect. McCone had been arguing for some time that they were, and one of his purposes at this meeting was to persuade Gates to join him in putting pressure on the President to end his policy of continuing the de facto moratorium. I said that I had just gone over every shred of intelligence we had on this matter and had found no evidence whatsoever supporting such a claim. McCone replied that my saying this was tantamount to treason. I was flustered by that awful charge, but I reasserted by position. Although the meeting ended inconclusively, I have never forgotten it. I believe that history fully confirms I was right. When the Soviets unilaterally ended the moratorium, they did so with a huge bang, not a secretive whimper. They did deceive us about their preparations during the months they needed to get their extensive test series ready, but there remains to this day no reason to believe they conducted any clandestine tests before they suddenly did so openly.

During my years at the Pentagon I maintained a relationship with the White House that was very unusual for someone in my position in the defense hierarchy. My calendar and the diary of George Kistiakowsky, the President's science adviser, show that I visited the White House and the old Executive Office Building about once a week over a three-year period. All of this was done with the full knowledge of the three Secretaries of Defense I worked for in succession: Neil McElroy, Tom Gates and Robert McNamara. I had a superb working relationship with all of them. They trusted me to deal directly with the White House on my own, and I behaved in accordance with that trust.

Among other things, I frequently acted as the de facto link between the Office of the Secretary of Defense on a number of matters. These included both intelligence about Soviet technical developments and the use of technology for producing such intelligence. The first included the Soviet nuclear program and missile program. The latter included such things as the programming of the U-2 "spy planes," the development of the new reconnaissance satellites and the monitoring of the nuclear test moratorium.

One of the issues that absorbed much of my attention throughout my years in Washington was the so-called missile gap. I first became in-

volved with it a few months before Sputnik as a result of my service on
the Gaither panel in the summer of 1957. That committee, headed by
Rowan Gaither, a San Francisco attorney and friend of Lawrence, took an
extreme view of the situation. It was not alone. That fall, after Sputnik,
Secretary of Defense McElroy and many others began to talk about a
coming "gap" favoring the Soviets, and the term *missile gap* entered political discourse with great force and frequency.

After I joined PSAC, and even more after I became a defense official, I
learned how we obtained the intelligence we had, and I also became more
sophisticated in interpreting it. A large part of what we knew then came
from the flights of the U-2 over the USSR, particularly in the region of
the missile test range near Tyuratam, in Kazakhstan. The more flights we
made, the more we learned about the Soviet development program. And
what we learned gave us pause. The Soviets had in fact successfully accomplished the development of a huge ICBM, several times as big as the
Atlas and Titan we were working on. It was capable of delivering very
large and powerful thermonuclear warheads to targets in our country in
less than 30 minutes. But we saw no deployments of such rockets. The
only launch facilities we found were those at Tyuratam. We continued to
search and came up with nothing.

Many of those privy to these facts, including the President, began to
suspect that the most probable explanation of why we found no rockets
was that there were none, or at most only a few. Gates and I came to share
this view. At the same time, important Democrats in the Congress, especially those eyeing the forthcoming 1960 Presidential elections, became
steadily more strident in their insistence on the reality and importance of
the missile gap. It was a difficult situation. We on the inside grew more
convinced that there was no missile gap, but we couldn't prove it, not
even to ourselves. The Soviet Union was, after all, the largest country on
Earth, 7 million square miles in all, and the U-2 flights had covered only a
tiny fraction of it. We were in effect trying to prove a negative on the basis of a very small sample.

In the spring of 1960 I made a calculation of the probable outcome of
an attack by Soviet ICBMs on our strategic forces at various times in the
near future. I assumed total surprise and used only the official CIA estimates for the numbers, accuracy, destructive power and reliability of Soviet missiles. The estimates of Soviet capability were reduced from their earlier "missile gap" values, but they remained formidable. I had doubts about
some of the figures, but I set them aside for the purpose of this calculation.

Our retaliatory forces then consisted of long-range bombers and
ICBMs based in North America, plus Thors and Jupiters based in Europe.

Polaris missiles were not scheduled to enter the force until late 1961.

Given all these initial assumptions, my calculations showed that for a period beginning earlier in 1961 and lasting for many months the Soviets could hypothetically reduce our retaliatory forces to zero in a surprise attack. I didn't call it that, but the situation was identical to the one that in later years would be referred to as the "window of vulnerability."

I also worked out a straightforward means for mitigating the problem. If we were to speed up construction of the Ballistic Missile Early Warning System, move the deployment of the first two Polaris submarines forward six months and arrange to put some of our bombers on air alert, the problem indicated by my calculations would be solved. With those changes, at no time could a Soviet attack bring our retaliatory forces down to zero. All this, of course, assumed that the CIA's estimates were right. Years later we learned that the estimates of Soviet force I used were still too high. There was therefore no such "window," but I could not know that at the time.

I briefed the Secretary of Defense and others, including the Joint Chiefs of Staff. The chiefs took the calculations seriously and urged me to bring them to the attention of President Eisenhower and the National Security Council. It was arranged that I should do so at an NSC meeting scheduled for the morning of 5 May 1960, to be held as usual in the Cabinet Room.

The afternoon before the meeting someone on the White House staff called and told me that instead of going directly to the White House the next morning, I should report to the Pentagon helicopter pad and bring my briefing charts with me. I did as directed. Herbert ("Pete") Scoville, then deputy director of the CIA, met me at the pad. We flew off toward the Blue Ridge Mountains of Virginia and landed near the entrance to one of the President's special remote underground command posts. Shortly thereafter the President came in another helicopter and the secretaries of Defense and State in a third. Only after the meeting did I learn that the regular members of the NSC, excepting only the President himself, had not been notified of the change in location until 6 o'clock in the morning of the day it was held. It was a genuine "fire drill" in the sense that the principals were taken by surprise.

Despite the unusual nature of the meeting, it proceeded in straightforward fashion. I gave a brief report of my calculations, which those present took seriously. Out of that meeting eventually came a plan to accelerate the Ballistic Missile Early Warning System and Polaris, and to arrange to place a fraction of our B-52s on air alert at such time as a future estimate might confirm the need.

President Eisenhower's farewell address is justly famed for its twin warnings about the "military-industrial complex" and the "scientific-technological elite." I was not surprised by his remarks, but, like many others, I wanted to know more about them. I had the opportunity to do so just a few years later. After leaving the presidency, Eisenhower spent his winters in Palm Desert, California, a town less than 100 miles from my home in La Jolla, and I called on him there on several occasions to pay my respects. Our conversation sometimes turned to the two warnings. I asked him to explain more fully what he meant, but he declined to do so, saying he didn't mean anything more detailed than what he had said at the time. I understood what he meant: The warnings were not the result of a methodical analysis; rather, they were the product of a remarkable intuition, whose power has generally been underestimated.

What, then, was the context of these remarks? What annoyed and irritated him? Whom are we to be wary of?

The context spanned the 40 months from the launching of Sputnik to the end of his administration. The people who irritated him were the hard-sell technologists who tried to exploit Sputnik and the missile gap psychosis it engendered. We were to be wary of accepting their claims, believing their analyses and buying their wares.

It seemed that the pursuit of extensive and complicated technology as an end in itself might become an accepted part of America's way of life. The Eisenhower administration, with the help of PSAC, was able to deal successfully and sensibly with most of the resulting rush of wild ideas, phony intelligence and hard sell.

At the beginning of the Eisenhower administration the authorities were hopeful that they could build air defenses capable of defending North America. We vigorously pursued the development of the means for doing so. These included radar warning networks, surface-to-air missiles of several kinds, and interceptor aircraft armed with a variety of antiair weapons. We also elaborated ideas and developed plans for vast civil defense programs. By the end of the Eisenhower administration, we concluded that we did not then know how to build a defense against missiles. We also decided, in effect, to stop extending our air defense system and to let what we had wither away. In sum, we decided to protect ourselves solely with the threat of nuclear retaliation. That policy has continued ever since.

Eisenhower considered the numbers of offensive weapons excessive, and believed that the quest for nuclear peace must be bolstered by diplomacy and negotiations of various kinds. He explored a number of possibilities—"atoms for peace," eliminating the threat of surprise attack and a nuclear test ban. By the time he left office, the International Atomic Energy Agency had become a reality, and a nuclear test moratorium was in place. The goal of the IAEA was to prevent the proliferation of nuclear weapons while promoting the broadest possible use of nuclear power for electricity. The test moratorium was intended as a first step toward wider and more important forms of arms limitation. Both succeeded, though not to the extent we had hoped. The current limited test ban and the other limitations on weapons that have been elaborated since then all derived directly from those first, modest successes.

Thus the three major elements that make up the current nuclear situation all assumed their present shape in the Eisenhower administration. These are (1) the number and types of weapons making up our strategic nuclear forces, including the concept of the triad (bombers, missiles and submarines); (2) the abandonment of any serious attempt to build either active defenses or civil defense, and the concomitant acceptance of a policy of maintaining nuclear peace through the threat of massive retaliation; and (3) the initiation of a search for political measures, including arms control, as a means of reinforcing and extending the quest for peace. In each of these areas the Eisenhower administration saw revolutionary changes. Since then, all further developments have been evolutionary. In 1983 President Reagan initiated the Strategic Defense Initiative, quickly dubbed Star Wars, whose basic purpose was to reverse number 2 above and diminish the importance of number 3. If he were to achieve these ends, it would bring about a new strategic revolution, but I am certain he will not.

During the 1960 Presidential campaign, a Democratic party study committee had recommended the creation of an Arms Control and Disarmament Agency. John F. Kennedy incorporated the idea into his platform as a candidate. After his election he made good on his campaign promise. The Arms Control Act of 1961, among other things, called for the creation of the General Advisory Committee to work with both the President and the agency director. Because of the politically delicate nature of arms control and disarmament, Congress insisted that its members be appointed by the President with the advice and consent of the Senate. They feared that otherwise the committee and the process might somehow be taken over by "woolly-headed" types. Senators opposing arms control were afraid such people would "give away the store." Senators favoring arms control were afraid they would give the process a bad name and cause the public to draw back from it.

John McCloy, who had been Kennedy's special assistant for that subject, was named chairman of the GAC. I became one of the members, and so did some other longtime friends and colleagues, including George Kistiakowsky and I.I. Rabi. I was chosen because I was thought to be both an expert on nuclear weapons and interested in pursuing arms control as an essential element of national security policy. Kistiakowsky and Rabi had similar reputations in this regard.

Our committee brought together a wide spectrum of views. One example involved B-47 bombers. The US was decommissioning B-47s as fast as it added ICBMs to its arsenal. At first we mothballed the airplanes, but it soon became evident that we should simply destroy them. Someone proposed that we do so in a big public bonfire celebrating the event. Two of the GAC members, representing opposite political extremes, opposed the idea.

"No. It would be phony and misleading," said Rabi, noting that we were replacing them with something better.

"No. We might need them again," said another member.

We also reviewed related programs in other agencies, particularly the Defense Department and CIA. One was Project Vela, a program I had been instrumental in starting when I was in ARPA. It was in essence a collection of technical devices and activities whose purpose was to monitor nuclear testing in all environments, including underground and in outer space.

Among those instruments was the so-called Vela satellite, designed to detect nuclear explosions in both space and atmosphere. In space it did so by detecting gamma rays emitted during nuclear explosions. As a happy by-product, the Vela satellite led to a whole new science—gamma-ray astronomy. In the atmosphere, the Vela satellite detected nuclear explosions by observing the intense light produced during the fireball stage of the explosion. Years later, in 1979, when I was in Geneva negotiating a test ban treaty with the British and the Soviets, one of these satellites detected a peculiar, intense light flash over the ocean south of Africa. The light signal was similar to, but not identical with, that emitted by a nuclear explosion. It proved to be unique in the 20-year history of the satellite. No other data corroborating a test at that time and place have ever turned up. Almost certainly it was a false alarm, but we could not know that when it was first reported. It became, for a while, the focus of much of our informal discussion in Geneva.

The two topics that interested me the most in those early days of ACDA were the nuclear test ban and the proposals to eliminate or limit antiballistic missiles or ABMs. Support for a test ban had been coming from two distinct groups: First, those who were primarily concerned about radioactive fallout and its harmful effects on human health and genetics; second, those who were primarily concerned about the connection between nuclear testing and the nuclear arms race. By eliminating tests in the atmosphere, the Limited Test Ban Treaty of 1963 effectively satisfied the people in the first category. Those in the second continued to push for a complete ban, but they did not have enough weight to do so now that the environmentalists were satisfied and had dropped out. Government efforts to achieve a comprehensive test ban continued until 1981, including a period of two years in 1979-81, when I was the US chief negotiator. We came close at times during those 18 years, but we never quite made it.

In 1981 President Reagan determined that a comprehensive test ban was not in the best US interests. Since then certain public interest groups have continued to push for it, but within the government the matter has been dropped.

To my knowledge, the first government official to propose and seriously study an international agreement to limit ABMs was Jack Ruina of MIT, then director of ARPA. Ruina's idea originated from a half joking remark made by Jerome Wiesner, then Kennedy's science adviser. Jerry had said something to the effect that the only reason our people wanted to build an ABM was because the Soviets were building one. At the time the Soviets were installing an ABM system—named the Galosh by NATO authorities—around Moscow, and Khrushchev boasted about how good it was. He claimed it could shoot a fly out of the sky and to prove his point he reminded us of the very annoying series of Soviet firsts in space.

Following up on Wiesner's remark. Ruina explored the possibility of a formal agreement to ban or limit such weapons. As Jack saw it, the situation in which no ABMs were deployed on either side had brought about "a curious and unprecedented stability," deriving from two factors: First, the military balance was insensitive to the number and kind of offensive weapons in the arsenals of each country so long as these were invulnerable; second, the danger that either side would miscalculate the consequences of a nuclear attack was minimized. The introduction of ABMs by either or both sides would change that by introducing new, important, but incalculable changes in the strategic relationship. These uncertainties, in turn, could lead both to an arms race instability—that is, to an unrestrained series of attempts by each side to cope with the worst possible case presented by the other side—and to instability at a time of crisis that could raise the pressure to go first.

The idea of limiting defenses seemed strange—indeed, even perverse—not only to the Russians when they first heard about it but to most members of our own defense establishments. Eventually, however, many high officials in both the US and the USSR, including Secretary of Defense McNamara, accepted the idea that limiting ABMs could provide an effective damper on the arms race.

In January 1967 McNamara carried the debate about ABMs into the Cabinet Room. He arranged a meeting at which, in addition to President Johnson, there were present all past and current special assistants to the President for science and technology—James Killian, Kistiakowsky, Wiesner and Donald Hornig—as well as all past and current Defense Department research directors—myself, Harold Brown and John Foster. We were asked the simple question that must be faced after all the complicated *if*'s, *and*'s, and *but*'s have been discussed: "Will it work and should it be deployed?" The outside experts all gave the same answer: "No. There is no prospect of its defending our people against a Soviet missile attack." McNamara said he would speak for the current Pentagon officials. To no one's surprise, he agreed with us outsiders. No one there contradicted him. It was my impression that Harold did in fact agree with him and that Johnny and the Joint Chiefs did not, but none of them were invited to give their views during that meeting.

According to Wiesner, a proposal to reappoint me to PSAC for a four-year term beginning January 1964 was on Kennedy's desk when he was assassinated. In that, as in most other matters during his first year in office, Johnson was true to his promise to continue what Kennedy had started. I was pleased to accept. The invitation came only weeks after my resignation as Chancellor of the University of California at San Diego, and I looked forward to spending a substantial effort on this new assignment. It turned out to be very different from what I had anticipated. The relations between the President and PSAC, particularly with its chairman, Hornig, were very different from those that had prevailed in the Killian, Kistiakowsky and Wiesner eras. A golden age of science advice to the President had come and gone in the span of only seven years.

Two meetings of the committee with President Johnson illustrate this point perfectly. The first took place soon after I rejoined the group. The President was open, cordial and, above all, optimistic. He said, "You just tell me what it is I should do. Don't you worry about how to get it done. That's my job and I'll take care of it." Three years later, toward the end of his term (and mine), we had another such meeting. He started that one by saying, "You people just come in here and tell me what I ought to do. You never stop to think how hard it is for me to do it, and you never take the time to help me with it."

Ever since I left full-time employment in the Pentagon in 1961, I have maintained a close working relationship with two special national security organizations—the Aerospace Corporation and the Institute for Defense Analyses, known simply as IDA. Each works mainly for one primary customer: Aerospace for the Department of the Air Force and IDA for the Office of the Secretary of Defense, including the Office of the Joint Chiefs of Staff. One of IDA's offspring is named Jason, which didn't leave home for the first 14 years of its life. Since then Jason has been successively a division of Stanford Research Institute and the Mitre Corporation. Jason is remarkable in many ways.

My involvement in starting Jason, combined with my continuing membership on the IDA board, made it natural for me to participate in many of Jason's activities, including its summer studies, even though I did not fit the standard profile of a Jason member. Many of the chiefs and leaders of Jason have been friends from my early Berkeley days: Keith Brueckner, Marvin Goldberger, Harold Lewis and William Nierenberg. Many other Jasons, including Dick Garwin, Sid Drell and Freeman Dyson, are people with whom I have had other long working relationships. Jason pioneered the work in beam weapons of all kinds. It studied a wide variety of strategic defense issues, ranging from basing modes for the MX to the interaction between nuclear explosions and detection systems. More recently, it has reviewed essentially all of the technical questions relevant to President Reagan's Star Wars proposal.

Jason also did pioneering work in arms control. One of the earliest instances involved the Multiple Independently Targeted Reentry Vehicle. In the 1964 Jason summer study, a group chaired by Ruina and including Dyson and Murray Gell-Mann examined the possible impact of new technologies on national security. MIRV was among them. The group concluded that its introduction, combined with foreseeable improvements in accuracy, would create a situation in which striking first could confer—or seem to confer—a substantial, perhaps decisive, advantage. The group was right on target. These developments were precisely what led to the "window of vulnerability" debate of the late 1970s and early 1980s.

The most controversial of Jason's many projects involved the "electronic battlefield" in Vietnam. The basic idea, which from the start received strong support from McNamara, called for installing a variety of special sensors in the jungles of Vietnam. The sensors would detect and report the presence of people or vehicles in the jungles and swamps of that unfortunate country. The main purpose was to deny the protection of natural cover to attacking enemy soldiers or infiltrating guerillas. Many organizations were involved in the concept, but Jason was central.

Jason's role in the project became known to the public. Because many students and professors were actively hostile to the war in Vietnam, and nearly all Jasons were college professors, a very dicey situation developed. Many Jasons, particularly those at Columbia University, were hassled and picketed by their students and colleagues. Others, including

Drell, were prevented from speaking at European universities, even on subjects having no relation to defense work. A few of the Jasons became disaffected with the war and dropped out of the group. Others simply stopped participating.

I, too, felt pressure—from both friends and my own conscience—to resign from Jason. I believed that the war was a bad mistake, that our cause was hopeless and that by continuing to fight we would only prolong the misery and increase the death and destruction. I did not, however, in any sense condone the North Vietnamese actions, as did many others on campuses and elsewhere, nor did I think it was a moral issue except in the general sense that all wars, especially modern ones, involve important moral questions and deep ethical contradictions.

More important, and despite my almost total lack of empathy with the main action going on, I continued to believe that the security of the US— and thus of the West—was a most worthy goal. I still continue my remaining relationships with the defense establishment, including Jason.

Many loyal and patriotic people did otherwise. Kistiakowsky decided it was too much for him. As a result, he publicly refused to have anything to do with the US defense establishment for the rest of his life. Some other colleagues also dropped out but without making any public statement. They simply ended their participation and declined any further invitations to give advice. In so behaving, they joined another group of veterans of defense science and technology—including Philip Morrison, Victor Weisskopf, Robert Wilson and many other Manhattan Project physicists who much earlier, in the first postwar years, decided they had done enough, or more than enough, of that kind of work.

Late in 1975 I received an invitation to a small cocktail reception from a local Democratic group. The words "JIMMY CARTER" were written in large, bright red letters across the top of the sheet. I had never heard of him, and I did not bother to attend. A year later he was elected President. By that time I was aware he had been a member of the Trilateral Commission and, through that connection, had become acquainted with several of my old friends and colleagues, including Harold Brown and Cyrus Vance.

I showed up at Brown's office a day or two after he became Secretary of Defense. He was reading a thick briefing book presenting the insiders' view of the history of his office and the permanent staff's ideas about the Secretary's functions. He was more elated than I can recall ever seeing him. It was evident that he would throw himself into the job with an energy and intensity that few others can bring to such responsibilities. I told him I wanted to help in whatever way I could. The result was another four years of full-time work on national security problems at the highest levels.

Both the form and the content of my contributions evolved over the four years. At first I worked as a direct consultant to Harold on issues having to do both with high-technology armaments and with arms control. At the end I served as chief US negotiator at the Comprehensive Test Ban talks in Geneva and otherwise dealt almost exclusively with arms control issues.

Since the bombing of Hiroshima all American presidents had actively sought means to contain, stop and reverse the nuclear arms race. Jimmy Carter differed from his predecessors not in kind but only in degree. He tried harder than they to explore the broadest set of possibilities. In one of modern history's all too common ironies, he accomplished the least. Events over which he had little control ultimately prevented him from adding very much to the limitations already worked out and put into place by the efforts of Eisenhower, Kennedy, Johnson, Nixon and Ford.

The principal White House study was directed by Carter's science adviser, Frank Press, a geophysicist from MIT. Press assembled a panel of experts including Bethe, Carson Mark, Wolfgang K.H. Panofsky, Ruina and me. In addition, the weapons laboratory directors, Harold Agnew of Los Alamos and Roger Batzel of Livermore, sat regularly with the panel. By that time, spring 1977, the argument over the utility of a test ban had come down to making a judgment about the relative value of two quite different factors, one of which weighed in on each side.

The main argument in favor of a test ban was that it was a necessary element of our nonproliferation policy. The Nonproliferation Treaty of 1970 called on the nuclear weapons powers to negotiate in good faith to end the nuclear arms race, and that was widely interpreted to require a comprehensive test ban as an early step. In 1975 the first quinquennial review of the nonproliferation regime focused attention on this point. To be sure, nuclear proliferation had proceeded much more slowly than had originally been expected, but this situation could change quickly. A change was especially likely if the superpowers continued to engage in "vertical proliferation," a phrase meaning the development and deployment of ever more varieties of nuclear weapons in their own arsenals.

The main argument against a test ban revolved around the issue of stockpile reliability. Nuclear weapons experts pointed out that these devices were built of both chemically active and radioactive materials, which steadily undergo changes than can adversely effect their performance. Occasional full-scale nuclear tests would be necessary in order to assure that old weapons still worked. Even rebuilt weapons would inevitably include small, supposedly harmless changes in their manufacture, and these, too, would have to be subjected to full-scale tests to assure performance. Opponents of a test ban also argued that we needed to continue testing in order to build safer and more secure bombs, to develop bombs properly optimized for new delivery systems and to learn more about weapons effects. Perhaps more important, continued testing was said to be needed in order to preserve a cadre of weapons design experts at the laboratories.

We studied the problem of stockpile reliability thoroughly and, except for the lab directors, decided that the nuclear establishment's worries were exaggerated. We concluded that regular inspections and nonnuclear tests of stockpiled bombs would uncover most such problems and provide solutions to them. Moreover, the laboratories could, if they tried, find ways around those that might remain. Agnew and Batzel disagreed. The in-house staffs in the Department of Energy and the Defense Nuclear Agency concurred with the laboratory directors, and the higher authorities in those agencies accepted their advice in the matter. The Joint Chiefs of Staff, whose nuclear arm is the Defense Nuclear Agency, also accepted the conclusions of the working-level experts immediately responsible for such matters. They really had no other choice.

Energy Secretary James Schlesinger also felt that President Carter was making a serious error in pushing ahead with a comprehensive test ban. Schlesinger had previously been chairman of the AEC, director of the CIA and Secretary of Defense, and his views were based on his experiences in those posts. In an attempt to dissuade Carter, he arranged to have the President meet with Agnew and Batzel so that they could explain to him why further tests were needed.

The intervention of the laboratory directors eventually caused a stir in the University of California—a stir that persisted for many years. The regents of the university managed the laboratories, and the directors were responsible to them through the university president. Most faculty members favored a test ban. In fact, roughly half of the faculty argued that the university should not be operating such labs. It was therefore no surprise that many faculty reacted negatively when they learned about what they regarded as unwarranted political intervention by persons who were, ostensibly, representing the university.

In sum, the Press panel reconfirmed Carter's intuitive view that a comprehensive test ban was in the national interest and could be adequately monitored. The strong opposition of the military and the nuclear establishment, however, made him realize that a test ban would be much more politically difficult to attain than a limitation on strategic forces.

Arms control has been a disappointment but not a failure. Large areas of the Earth, including Latin America and Antarctica, remain free of nuclear weapons at least partly because of deliberate diplomatic efforts to keep them that way. The most successful, and arguably the most important, of all arms control policies have been those designed to limit the spread or proliferation of nuclear weapons to additional states.

Despite almost universal expectations to the contrary, there are still only five overt nuclear powers, and the last state to become one, China, did so in 1964, a full generation ago. In the quarter of a century since that happened, only four others—India, Israel, Pakistan and South Africa—have taken strong steps toward becoming nuclear powers, but none of them has yet built and deployed substantial, overt nuclear forces. This surprising but happy result must be credited to the combined antiproliferation policies and actions of the majority of the world's states.

After 30 years of actively working for a comprehensive test ban, I have been forced to conclude, as I did in the 1960s, that it will be politically possible and stable only in a world in which the great powers are clearly and forcefully moving away from their current dependence on nuclear weapons.

I crystallized my thinking about the correct US approach to national security early in 1961 as I prepared to meet with John McCloy, who had just been appointed by President Kennedy as his arms control and disarmament adviser and emissary. I organized my thoughts around three basic principles, each derived from my experiences of the last several years: (1) defense of the population is impossible in the nuclear era, (2) our national security dilemma has no technical solution and (3) our only real hope for the long run lies in working out a political solution.

The notion that we must "do something" radical soon about either the Soviet threat or the nuclear arms race or be doomed has been with us for more than 40 years. So far it has always proved to be wrong, and I expect it will remain so for the foreseeable future. The maintenance of an adequate balance of power, including the nuclear component, combined with classical diplomatic actions designed to control arms and preserve the peace, has bought us time. If we are wise enough, we will use it to find a way out of the grand nuclear dilemma.

1994 Postscript

The last paragraphs of the foregoing essay, written in 1987, contain two references to future possibilities. One concerns the conditions for achieving a nuclear test ban, the other concerns peace and stability more generally. The events of the intervening years promise to turn those possibilities into realities.

We have indeed arrived at "a world in which the great powers are clearly and forcefully moving away from their current dependencies on nuclear weapons" and a real comprehensive test ban is in sight. The collapse of the Soviet Union, which no one foresaw when I wrote this essay, led directly to a series of deep reductions in arms, some negotiated and some simply unilaterally announced. By 1994 these actions and announcements had resulted in major reductions in the numbers of weapons, and most of those that remain have either been moved from their former forward deployment to some sort of central storage or otherwise had their state of readiness substantially reduced. If all the plans and promises announced so far are actually fulfilled, the result will be a net ten fold reduction in numbers.

These changes in numbers have been accompanied by, and are in fact partly due to, a real "sea change" in the way defense officials look at nuclear weapons. Where once nuclear weapons were viewed primarily as a solution to certain otherwise intractable security problems, they are now coming to be seen primarily as a problem in themselves.

One of the concrete results of all these changes is a revival in the prospects for a comprehensive test ban. At the time of this writing, a moratorium on testing is in effect, and test ban negotiations are again underway in Geneva. Much remains to be worked out in this regard, including especially the agreement of all the other nuclear weapons states, but the prospects have never before looked so promising.

The last lines in this essay express the hope that our long-term nuclear policy, instituted by Truman and Eisenhower and continued by every subsequent president, a policy which I characterize as one of *making weapons while talking peace*, will buy us the time we need to work our way out of the grand nuclear dilemma we stumbled into at the beginning of the Cold War. It is still a little too soon to be certain, but that may be what is in fact happening. The changes made so far do not yet quite free us from the grasp of that dilemma, but mainstream intellectuals and statesmen in the major countries are beginning to think seriously about proposals that would do so. Many of these proposals were widely regarded as too radical only a few years ago; they include further reductions in stockpiles to only a few hundred in the foreseeable future, and perhaps even to zero eventually.

ARMS AND INSECURITY

National Security and the Nuclear-Test Ban

WITH JEROME B. WIESNER

The partial nuclear-test ban—the international treaty that prohibits nuclear explosions in the atmosphere, in the oceans and in outer space—went into effect in 1963. From July, 1945, when the first atomic bomb was set off in New Mexico, until August, 1963, when the U.S. completed its last series of atmospheric bomb tests in the Pacific, the accumulated tonnage of nuclear explosions had been doubling every three years. Contamination of the atmosphere by fission products and by the secondary products of irradiation (notably the long-lived carbon 14) was approaching a level (nearly 10 percent of the natural background radiation) that alarmed many biologists. A chart plotting the accumulation of radioactive products can also be read as a chart of the acceleration in the arms race.

Now that curve has flattened out. From the objective record it can be said that within that time the improvement of both the physical and the political atmosphere of the world has fulfilled at least the short-range expectations of those who advocated and worked for the test ban. In and of itself the treaty does no more than moderate the continuing arms race. It is nonetheless, as President Kennedy said, "an important first step—a step toward peace, a step toward reason, a step away from war."

The passage of a year also makes it possible to place in perspective and evaluate certain misgivings that have been expressed about the effect on U.S. national security of the suspension of the testing of nuclear weapons in the atmosphere. These misgivings principally involve the technology of nuclear armament. National security, of course, involves moral questions and human values—political, social, economic and psychological questions as well as technological ones. Since no one is an expert in all the disciplines of knowledge concerned, it is necessary to consider one class of such questions at a time, always with the caution that such consideration is incomplete. As scientists who have been engaged for most of our professional lifetimes in consultation on this country's military policy and in the active development of the weapons themselves, we shall devote the present discussion primarily to the technological questions.

The discussion will necessarily rest on unclassified information. It is unfortunate that so many of the facts concerning this most important problem are classified, but that is the situation at this time. Since we have access to classified information, however, we can assure the reader that we would not have to modify any of the arguments we present here if we were able to cite such information. Nor do we know of any military considerations excluded from open discussion by military secrecy that would weaken any of our conclusions. We shall discuss the matter from the point of view of our country's national interest. We believe, however, that a Soviet military technologist, writing from the point of view of the U.S.S.R., could write an almost identical paper.

Today as never before national security involves technical questions. The past two decades have seen a historic revolution in the technology of war. From the blockbuster of World War II to the thermonuclear bomb the violence of military explosives has been scaled upward a million times. The time required for the interhemispheric transport of weapons of mass destruction has shrunk from 20 hours for the 300-mile-per-hour B-29 to the 30-minute flight time of the ballistic missile. Moreover, the installation of the computer in command and control systems has increased their information-processing capacity by as much as six orders of magnitude compared with organizations manned at corresponding points by human nervous systems.

It has been suggested by some that technological surprise presents the primary danger to national security. Yet recognition of the facts of the present state of military technology must lead to the opposite conclusion. Intercontinental delivery time cannot be reduced to secure any significant improvement in the effectiveness of the attack. Improvement by another order of magnitude in the information-processing capacity of the defending system will not make nearly as large a difference in its operational effectiveness.

The point is well illustrated by the 100-megaton nuclear bomb. Whether or not it is necessary, in the interests of national security, to test and deploy a bomb with a yield in the range of 100 megatons was much discussed during the test-ban debates. The bomb was frequently referred to as the "big" bomb, as if the bombs now in the U.S. arsenal were somehow not big. The absurdity of this notion is almost enough by itself to settle the argument. A one-megaton bomb is already about 50 times bigger than the bomb that produced 100,000 casualties at Hiroshima, and 10 megatons is of the same order of magnitude as the grand total of all high explosives used in all wars to date. Other technical considerations that surround this question are nonetheless illuminating and worth exploring.

There is, first of all, the "tactics" of the missile race. The purpose of a missile system is to be able to destroy or, perhaps more accurately, able to threaten to destroy enemy targets. No matter what the statesmen, military men and moralists on each side may think of the national characteristics, capabilities and morality of the other side, no matter what arguments may be made about who is aggressive and who is not or who is rational and who is not, the military planners on each side must reckon with the possibility that the other side will attack first. This means that above all else the planner must assure the survival of a sufficient proportion of his own force, following the heaviest surprise attack the other side might mount, to launch a retaliatory attack. Moreover, if the force is to be effective as a deterrent to a first strike, its capacity to survive and wreak revenge and even win, whatever that may mean, must be apparent to the other side.

Several approaches, in fact, can be taken to assure the survival of a sufficient missile force after a first attack on it. The most practical of these are: (1) "hardening," that is, direct protection against physical damage; (2) concealment, including subterfuge and, as in the case of the Polaris submarine missiles, mobility, and (3) numbers, that is, presenting more targets than the attacker can possibly cope with. The most straightforward and certain of these is the last: numbers. For the wealthier adversary it is also the easiest, because he can attain absolute superiority in numbers. A large number of weapons is also a good tactic for the poorer adversary, because numbers even in the absence of absolute superiority can hopelessly frustrate efforts to locate all targets.

There is an unavoidable trade-off, however, between the number and the size of weapons. The cost of a missile depends on many factors, one of the most important being gross size or weight. Unless one stretches "the state of the art" too far in the direction of sophistication and miniaturization, the cost of a missile turns out to be roughly proportional to its weight, if otherwise identical design criteria are used. The protective structures needed for hardening or the capacity of submarines needed to carry the missile also have a cost roughly proportional to the volume of the missile. Some of the ancillary equipment has a cost proportional to the size of the missile and some does not; some operational expenditures vary directly with size or weight and some do not. The cost of the warhead generally does not, although the more powerful warhead requires the larger missile. It is not possible to put all these factors together in precise bookkeeping form, but it is correct to say that the cost of a missile, complete and ready for firing, increases somewhat more slowly than linearly with its size.

On the other hand—considering "hard" targets only—the effectiveness of a missile increases more slowly than cost as the size of the missile goes up. The reason is that the radius of blast damage, which is the primary effect employed against a hard target, increases only as the cube root of the yield and because yield has a more or less direct relation to weight. Against "soft" targets, meaning population centers and conventional military bases, even "small" bombs are completely effective, and nothing is gained by increasing yield. Given finite resources, even in the wealthiest economy, it would seem prudent to accept smaller size in order to get larger numbers. On any scale of investment, in fact, the combination of larger numbers and smaller size results in greater effectiveness for the missile system as a whole, as contrasted to the effectiveness of a single missile.

This line of reasoning formed the basis of U.S. missile policy. The administration of President Eisenhower, when faced with the choice of bigger missiles (the liquid-fueled Atlas and Titan rockets) as against smaller missiles (the solid-fueled Minuteman and Polaris rockets), decided to produce many more of the smaller missiles. The administration of President Kennedy independently confirmed this decision and increased the ratio of smaller to larger missiles in the nation's armament. During the test-ban hearings it was revealed that the U.S. nuclear armament included bombs of 23-megaton yield and higher, carried by bombers. Recently Cyrus R. Vance, Under Secretary of Defense, indicated that the Air Force has been retiring these large bombs in favor of smaller ones. There are presumably no targets that call for the use of such enormous explosions.

The argument that says it is now critical in 1964 for U.S. national security to build very big bombs and missiles fails completely when it is examined in terms of the strictly technical factors that determine the effectiveness of a missile attack. In addition to explosive yield the principal factors are the number of missiles, the overall reliability of each missile and the accuracy with which it can be delivered to its target. The effectiveness of the attack—the likelihood that a given target will be destroyed—can be described by a number called the "kill probability" (P_k). This number depends on the number of missiles (N) launched at the target, the reliability (r) of each missile and the ratio of the radius of damage

(R_k) effected by each missile to the accuracy with which the missiles are delivered to the target (CEP). The term "CEP," which stands for "circular error probable," implies that the distribution of a large number of hits around a given target will follow a standard error curve. The term "CEP" is still useful, however, and can be defined simply as the circle within which half of a large number of identical missiles would fall.

Now, in the case of a soft target, R_k is very large for the present range of warhead yields in the U.S. arsenal. The reason is that soft targets are so highly vulnerable to all the "prompt" effects (particularly the incendiary effects) of thermonuclear weapons. The range of these effects, modified by various attenuation factors, increases approximately as the square root or the cube root of the yield at large distances. Under these circumstances, given the accuracy of existing fire-control systems, the ratio R_k/CEP is large and the likelihood that the target will be destroyed becomes practically independent of this ratio. Instead P_k depends primarily on r, the reliability of the missile. If r is near unity, then a single missile ($N = 1$) will do the job; if r is not near unity, then success in the attack calls for an offsetting increase in the number of missiles [$P_k = 1 - (1 - r)^N$]. In either case changes in R_k make little difference. That is to say, a "big" bomb cannot destroy a soft target any more surely than a "small" one can.

When it comes to hard targets, the ratio R_k/CEP becomes much smaller even for bombs of high yield. The blast effects—including the ground rupture, deformation and shock surrounding the crater of a surface burst—have comparatively small radii at intensities sufficient to overcome hardening. Moreover, as mentioned above, the radii of these effects increase only as the cube root of the yield. This rule of thumb is modified somewhat in both directions by the duration of the blast pulse, local variations in geology and other factors, but it is sustained by a voluminous record from weapons tests. Since the radius of blast damage is of the same order of size as the circular error probable, or smaller, the ratio R_k/CEP must be reckoned with in an attack on a hard target. Yet even in this situation the cube root of a given increase in yield would contribute much less to success than a comparable investment in numbers, reliability or accuracy.

Yield is of course a product of the yield-to-weight ratio of the nuclear explosive employed in the warhead multiplied by the weight of the warhead. In order to gain significant increases in the first of these two quantities further nuclear tests would be necessary. Increase in the weight of the warhead, on the other hand, calls for bigger and more efficient missiles. Efforts to improve CEP and reliability as well as weight-carrying capacity hold out more promise than efforts to improve the yield-to-weight ratio. The reason is that missile design and control involve less mature and less fully exploited technologies than the technology of nuclear warheads. Finally, an increase in the number of missiles, although not necessarily cheap, promises more straightforward and assured results than a fractional increase in yield-to-weight ratio. Of all the various possible technical approaches to improving the military effectiveness of an offensive missile force, therefore, the only one that calls for testing (whether underground or in the atmosphere) is the one that offers the smallest prospect of return.

Suppose, however, a new analysis, based on information not previously considered, should show that it is in fact necessary to incorporate the 100-megaton bomb in the U.S. arsenal. Can this be done without further weapons tests? The answer is yes. Because the U.S.S.R. has pushed development in this yield range and the U.S. has not, the U.S. 100-megaton bomb might not be as elegant as the Soviet model. It would perhaps weigh somewhat more or at the same weight would produce a somewhat lower yield. It could be made, however, and the basic techniques for making it have been known since the late 1950's. The warhead for such a bomb would require a big missile, but not so big as some being developed by the National Aeronautics and Space Administration for the U.S. space-exploration program. Such a weapon would be expensive, particularly on a per-unit basis; under any imaginable circumstances it would be of limited use and not many of its kind would be built.

The extensive series of weapons tests carried out by the U.S.—involving the detonation of several hundred nuclear bombs and devices—have yielded two important bodies of information. They have shown how to bring the country's nuclear striking force to its present state of high effectiveness. And they have demonstrated the effects of nuclear weapons over a wide range of yields. Among the many questions that call for soundly based knowledge of weapons effects perhaps none is more important in a discussion of the technical aspects of national security than: What would be the result of a surprise attack by missiles on the country's own missile forces? Obviously if the huge U.S. investment in its nuclear armament is to succeed in deterring an attacker, that armament must be capable of surviving a first strike.

A reliable knowledge of weapons effects is crucial to the making of rational decisions about the number of missiles needed, the hardening of missile emplacements, the degree of dispersal, the proportion that should be made mobile and so on. The military planner must bear in mind, however, that such decisions take time—years—to carry out and require large investments of finite physical and human resources. The inertia of the systems is such that the design engineer at work today must be concerned not with the surprise attack that might be launched today but rather with the kind and size of forces that might be launched against them years in the future. In addition to blast, shock and other physical effects, therefore, the planner must contend with a vast range of other considerations. These include the yields of the various bombs the attacker would use against each target; the reliability and accuracy of his missiles; the number and kind of weapons systems he would have available for attack; the tactics of the attacker, meaning the number of missiles he would commit to a first strike, the fractions he would allocate to military as against civilian targets and the relative importance he would assign to various kinds of military targets, the effects of chaos on the defender's capacity to respond, and so on. In all cases the planner must project his thinking forward to some hypothetical future time, making what he can of the available intelligence about the prospective attacker's present capabilities and intentions. Plainly all the "other considerations" involve inherently greater uncertainties than the knowledge of weapons effects.

The extensive classified and unclassified literature accumulated in two decades of weapons tests and available to U.S. military planners contains at least some observations on all important effects for weapons with a large range of yields. These observations are more or less well understood in terms of physical theories; they can be expressed in numerical or algebraic form, and they can be extrapolated into areas not fully explored in the weapons tests conducted by the U.S., for example into the 100-megaton range. As one departs from the precise circumstances of past experiments, of course, extrapolation becomes less and less reliable. Nonetheless, some sort of estimate can be made about what the prompt and direct effects will be under any conceivable set of circumstances.

Consider, in contrast, the degree of uncertainty implicit in predicting the number and kind of weapons systems that might be available to the prospective attacker. Such an uncertainty manifested itself in the famous "missile gap" controversy. The remarkable difference between the dire predictions made in the late 1950's—based as they were on the best available intelligence—and the actual situation that developed in the early 1960's can be taken as indicating the magnitude of the uncertainties that surround the variables other than weapons effects with which the military planner must contend. Moreover, these factors, as they concern a future attack, are uncertain not only to the defender; they are almost as uncertain to the attacker.

Uncertainties of this order and kind defy reduction to mathematical expression. A human activity as complex as modern war cannot be computed with the precision possible in manipulation of the data that concern weapons effects. What is more, the uncertainties about this single aspect of the total problem are not, as is sometimes assumed, multiplicative in estimation of the overall uncertainty. Most, but not all, of the uncertainties are independent of one another. The total uncertainty is therefore, crudely speaking, the square root of the sum of the squares of the individual uncertainties.

In our view further refinement of the remaining uncertainties in the data concerning prompt direct physical effects can contribute virtually nothing more to management of the real military and political problems, even though it would produce neater graphs. Furthermore, if new effects should be discovered either experimentally or theoretically in the future, or if, in certain peculiar environments, some of the now known effects should be excessively uncertain, it will be almost certainly possible to "over-design" the protection against them. Thus, although renewed atmospheric testing would contribute some refinement to the data on weapons effects, the information would be, at best, of marginal value.

Such refinements continue to be sought in the underground tests that are countenanced under the partial test ban. From this work may also come some reductions in the cost of weapons, modest improvements in yield-to-weight ratios, devices to fill in the spectrum of tactical nuclear weapons and so on. There is little else to justify the effort and expenditure. The program is said by some to be necessary, for example, to the development of a pure fusion bomb, sometimes referred to as the "neutron bomb." It is fortunate that this theoretically possible (stars are pure fusion systems) device has turned out to be so highly difficult to create; if it were relatively simple, its development might open the way to thermonuclear armament for the smallest and poorest powers in the world. The U.S., with its heavy investment in the fission-to-fusion technology, would be the last nation to welcome this development and ought to be the last to encourage it. Underground testing is also justified for its contribution to the potential peaceful uses of nuclear explosives. Promising as these may be, the world could forgo them for a time in exchange for cessation of the arms race. Perhaps the best rationale for the underground-test program is that it helps to keep the scientific laboratories of the military establishment intact and in readiness—in readiness, however, for a full-scale resumption of the arms race.

Paradoxically one of the potential destabilizing elements in the present nuclear standoff is the possibility that one of the rival powers might develop a successful antimissile defense. Such a system, truly airtight and in the exclusive possession of one of the powers, would effectively nullify the deterrent force of the other, exposing the latter to a first attack against which it could not retaliate. The possibilities in this quarter have often been cited in rationalization of the need for resuming nuclear tests in the atmosphere. Here two questions must be examined. One must first ask if it is possible to develop a successful antimissile defense system. It then becomes appropriate to consider whether or not nuclear weapons tests can make a significant contribution to such a development.

Any nation that commits itself to large-scale defense of its civilian population in the thermonuclear age must necessarily reckon with passive modes of defense (shelters) as well as active ones (antimissile missiles). It is in the active mode, however, that the hazard of technological surprise most often lurks. The hazard invites consideration if only for the deeper insight it provides into the contemporary revolution in the technology of war.

The primary strategic result of that revolution has been to overbalance the scales in favor of the attacker rather than the defender. During World War II interception of no more than 10 percent of the attacking force gave victory to the defending force in the Battle of Britain. Attrition of this magnitude was enough to halt the German attack because it meant that a given weapons-delivery system (bomber and crew) could deliver on the average only 10 payloads of high explosive; such a delivery rate was not sufficient to produce backbreaking damage. In warfare by thermonuclear missiles the situation is quantitatively and qualitatively different. It is easily possible for the offense to have in its possession and ready to launch a number of missiles that exceeds the number of important industrial targets to be attacked by, let us say, a factor of 10. Yet the successful delivery of only one warhead against each such target would result in what most people would consider an effective attack. Thus where an attrition rate of only 10 percent formerly crowned the defense with success, a penetration rate of only 10 percent (corresponding to an attrition rate of 90 percent) would give complete success to the offense. The ratio of these two ratios is 100 to one; in this sense the task of defense can be said to have become two orders of magnitude more difficult.

Beyond this summary statement of the situation there are many general reasons for believing that defense against thermonuclear attack is impossible. On the eve of attack the offense can take time to get ready and to "point up" its forces; the defense, meanwhile, must stay on alert over periods of years, perpetually ready and able to fire within the very few minutes available after the first early warning. The attacker can pick its targets and can choose to concentrate its forces on some and ignore others; the defense must be prepared to defend all possible important targets. The offense may attack the defense itself; then, as soon as one weapon gets through, the rest have a free ride.

The hopelessness of the task of defense is apparent even now in the stalemate of the arms race. A considerable inertia drags against the movement of modern, large-scale, unitary weapons systems from the stage of research and development to operational deployment. The duration and magnitude of these enterprises, whether defensive or offensive, practically assure that no system can reach full deployment under the mantle of secrecy. The designer of the defensive system, however, cannot begin until he has learned something about the properties and capabilities of the offensive system. Inevitably the defense must start the race a lap behind. In recent years, it seems, the offense has even gained somewhat in the speed with which it can put into operation stratagems and devices that nullify the most extraordinary achievements in the technology of defense. These general observations are expensively illustrated in the development and obsolescence of two major U.S. defense systems.

Early in the 1950's the U.S. set out to erect an impenetrable defense against a thermonuclear attack by bombers. The North American continent was to be ringed with a system of detectors that would flash information back through the communications network to a number of computers. The computers were to figure out from this data what was going on and what ought to be done about it and then flash a series of commands to the various interceptor systems. In addition to piloted aircraft, these included the Bomarc (a guided airborne missile) and the Nike-Hercules (a ballistic rocket). By the early 1960's this "Sage" system was to be ready to detect, intercept and destroy the heaviest attack that could be launched against it.

The early 1960's have come and yet nothing like the capability planned in the 1950's has been attained. Why not? Time scales stretched out, subsystems failed to attain their planned capabilities and costs increased. Most important, the offense against which the system was designed is not the offense that actually exists in the early 1960's. Today the offensive system on both sides is a mixture of missiles and bombers. The Sage system has a relatively small number of soft but vital organs completely vulnerable to missiles—a successful missile attack on them would give a free ride to the bombers. As early as 1958 the Department of Defense came to realize that this would be the situation, and the original grand plan was steadily cut back. In other words, the Sage system that could have been available, say, in 1963 and that should have remained useful at least through the 1960's would in principle have worked quite well against the offense that existed in the 1950's.

To answer the intercontinental ballistic missile, the Department of Defense launched the development of the Nike-Zeus system. Nike-Zeus was intended to provide not a defense of the continent at its perimeter but a point defense of specific targets. To be sure, the "points" were fairly large—the regions of population concentration around 50 to 70 of the country's biggest cities. The system was to detect incoming warheads, feeding the radar returns directly into its computers, and launch and guide an interceptor missile carrying a nuclear warhead into intersection with the trajectory of each of the incoming warheads.

Nike-Zeus was not designed to defend the 1,000 or so smaller centers outside the metropolitan areas simply because there are too many of these to be covered by the resources available for a system so huge and complicated. Nor was the system designed to defend the retaliatory missiles, the security of these forces being entrusted to the more reliable protection of dispersal, concealment, mobility and number. In principle, the defense of a hardened missile silo would have presented by far the simplest case for proof of the effectiveness of Nike-Zeus as advanced by those who contend that such a system can be made to "work." There would be no ambiguity about the location of the target of the incoming warhead. By the same token Nike-Zeus might have been considered for the defense of a few special defense posts, such as the headquarters of the Air Defense Command of the Strategic Air Command. These special cases are so few in number, however, that it had to be concluded that the attacker would either blast his way through to them by a concentration of firepower or ignore them altogether.

At the time of the conception of the Nike-Zeus system its designers were confronted with a comparatively simple problem, namely that of shooting down the warheads one by one as they presented themselves to the detectors. Even this simple problem had to be regarded as essentially unsolvable, in view of the fact that a 90 percent success in interception constitutes failure in the inverted terms of thermonuclear warfare. At first, therefore, the designers of the offensive system did not take the prospect of an antimissile system seriously. Then the possibility that the problem of missile interception might be solved in principle gave them pause. Thereupon the designers of the offense began to invent a family of "penetration aids," that is, decoys and confusion techniques. The details of these and the plans for their use are classified, but the underlying principles are obvious. They include light decoys that can be provided in large numbers but that soon betray their character as "atmospheric sorting" separates them from the heavier decoys (and actual warheads) that can be provided in smaller numbers to confuse the defending detectors down to the last minute. Single rockets can also eject multiple warheads. Both the decoys and the warheads can be made to present ambiguous cross sections to the radar systems. These devices and stratagems overwhelmed the designed capability of the Nike-Zeus system and compelled its recent abandonment.

If the installation of the system had proceeded according to plan, the first Nike-Zeus units would have been operational within the next year or two. This could have been celebrated as a technical milestone. As a means of defense of a substantial percentage of the population, however, the system would not have reached full operational deployment until the end of the decade. In view of its huge cost the system should then have looked forward to a decade of useful life until, say, the late 1970's. Thus, in inexorable accordance with the phase-lag of the defense, the U.S. population was to be defended a decade too late by a system that might have been effective in principle (although most probably not in practice) against the missiles of the early 1960's.

The race of the tortoise and the hare has now entered the next lap with the development of the Nike-X system as successor to Nike-Zeus. The Advanced Research Projects Agency of the Department of Defense has been spending something on the order of $200 million a year on its so-called Defender Program, exploring on the broadest front the principles and techniques that might prove useful in the attempt to solve the antimissile problem. Although nothing on the horizon suggests that there is a solution, this kind of work must go forward. It not only serves the forlorn hope of developing an active antimissile defense but also promotes the continued development of offensive weapons. The practical fact is that work on defensive systems turns out to be the best way to promote invention of the penetration aids that nullify them.

As the foregoing discussion makes clear, the problems of antimissile development are problems in radar, computer technology, missile propulsion, guidance and control. The nuclear warheads for the antimissile missile have been ready for a long time for delivery to the right place at the right time. Although it is argued that certain refinements in the existing data about weapons effects are needed, the other uncertainties all loom much larger than the marginal uncertainties in these physical effects. The antimissile defense problem, then, is one in which nuclear testing can play no really significant part.

The pursuit of an active defense system demands parallel effort on the passive defense, or shelter, front because the nature of the defense system strongly conditions the tactics of the offense that is likely to be mounted against it. To take a perhaps farfetched example, a Nike-Zeus system that provided protection for the major population centers might invite the attacker to concentrate the weight of his assault in ground bursts on remote military installations and unprotected areas adjacent to cities, relying on massive fallout to imperil the population centers. This example serves also to suggest how heavily the effectiveness of any program for sheltering the civilian population depends on the tactics of the attacker.

Fallout shelters by themselves are of no avail if the attacker chooses to assault the population centers directly.

In any speculation about the kind of attack to which this country might be exposed it is useful to note where the military targets are located. Most of the missile bases are, in fact, far from the largest cities. Other key military installations, however, are not so located. Boston, New York, Philadelphia, Seattle, San Francisco, Los Angeles (Long Beach) and San Diego all have important naval bases. Essential command and control centers are located in and near Denver, Omaha and Washington D.C. The roll call could be extended to include other major cities containing military installations that would almost certainly have to be attacked in any major assault on this country. The list does not stop with these; it is only prudent to suppose still other cities would come under attack, because there is no way to know in advance what the strategy may be.

The only kind of shelter that is being seriously considered these days, for other than certain key military installations, is the fallout shelter. By definition fallout shelters offer protection against nothing but fallout and provide virtually no protection against blast, fire storms and other direct effects. Some people have tried to calculate the percentage of the population that would be saved by fallout shelters in the event of massive attack. Such calculations always involve predictions about the form of the attack, but since the form is unknowable the calculations are nonsensical. Even for the people protected by fallout shelters the big problem is not a problem in the physical theory of gamma-ray attenuation, which can be neatly computed, but rather the sociological problem of the sudden initiation of general chaos, which is not subject to numerical analysis.

Suppose, in spite of all this, the country were to take fallout shelters seriously and build them in every city and town. The people living in metropolitan areas that qualify as targets because they contain essential military installations and the people living in metropolitan areas that might be targeted as a matter of deliberate policy would soon recognize that fallout shelters are inadequate. That conclusion would be reinforced by the inevitable reaction from the other side, whose military planners would be compelled to consider a massive civilian-shelter program as portending a first strike against them. Certainly the military planners of the U.S. would be remiss if they did not take similar note of a civilian-shelter program in the U.S.S.R. As a step in the escalation of the arms race toward the ultimate outbreak of war, the fallout shelter would lead inevitably to the blast shelter. Even with large numbers of blast shelters built and evenly distributed throughout the metropolitan community, people would soon realize that shelters alone are not enough. Accidental alarms, even in tautly disciplined military installations, have shown that people do not always take early warnings seriously. Even if they did, a 15-minute "early" warning provides less than enough time to seal the population into shelters. Accordingly, the logical next step is the live-in and work-in blast shelter leading to still further disruption and distortion of civilization. There is no logical termination of the line of reasoning that starts with belief in the usefulness of fallout shelters; the logic of this attempt to solve the problem of national security leads to a diverging series of ever more grotesque measures.

Ever since shortly after World War II the military power of the U.S. has been steadily increasing. Throughout this same period the national security of the U.S. has been rapidly and inexorably diminishing. In the early 1950's the U.S.S.R., on the basis of its own unilateral decision and determination to accept the inevitable retaliation, could have launched an attack against the U.S. with bombers carrying fission bombs. Some of these bombers would have penetrated our defenses and the American casualties would have numbered in the millions. In the later 1950's, again on its own sole decision and determination to accept the inevitable massive retaliation, the U.S.S.R. could have launched an attack against the U.S. using more and better bombers, this time carrying thermonuclear bombs. Some of these bombers would have penetrated our defenses and the American casualties could have numbered in the tens of millions.

Today the U.S.S.R., again on the basis of its own decision and determination to accept the inevitable retaliation, could launch an attack on the U.S. using intercontinental missiles and bombers carrying thermonuclear weapons. This time the number of American casualties could very well be on the order of 100 million.

The steady decrease in national security did not result from any inaction on the part of responsible U.S. military and civilian authorities. It resulted from the systematic exploitation of the products of modern science and technology by the U.S.S.R. The air defenses deployed by the U.S. during the 1950's would have reduced the number of casualties the country might have otherwise sustained, but their existence did not substantively modify this picture. Nor could it have been altered by any other defense measures that might have been taken but that for one reason or another were not taken.

From the Soviet point of view the picture is similar but much worse. The military power of the U.S.S.R. has been steadily increasing since it became an atomic power in 1949. Soviet national security, however, has been steadily decreasing. Hypothetically the U.S. could unilaterally decide to destroy the U.S.S.R. and the U.S.S.R. would be absolutely powerless to prevent it. That country could only, at best, seek to wreak revenge through whatever retaliatory capability it might then have left.

Both sides in the arms race are thus confronted by the dilemma of steadily increasing military power and steadily decreasing national security. *It is our considered professional judgment that this dilemma has no technical solution.* If the great powers continue to look for solutions in the area of science and technology only, the result will be to worsen the situation. The clearly predictable course of the arms race is a steady open spiral downward into oblivion.

We are optimistic, on the other hand, that there is a solution to this dilemma. The partial nuclear-test ban, we hope and believe, is truly an important first step toward finding a solution in an area where a solution may exist. A next logical step would be the conclusion of a comprehensive test ban such as that on which the great powers came close to agreement more than once during 10 long years of negotiation at Geneva. The policing and inspection procedures so nearly agreed on in those parleys would set significant precedents and lay the foundations of mutual confidence for proceeding thereafter to actual disarmament.

The Arms Race and the Fallacy
of the Last Move

I t is important to note that in hard-point defense, methods which cannot be used to defend cities or large areas become feasible. Such approaches include mobility (as in Polaris and Poseiden), deployment
of greater numbers of offensive missiles and various deception devices
and tactics such as providing more missile silo targets than there are missiles and then playing a sort of shell game with the missiles themselves.
Thus it is precisely in the case where an ABM-type defense becomes easiest that numbers of alternative technical defense schemes also become
possible. Furthermore, and again because the problem as given is easier, it
is quite safe to postpone any decision to deploy an ABM at least until after present attempts to get new arms control negotiations moving.

I should like now to turn to a technical problem that pertains to all the
forms of ABM so far proposed, but which unfortunately is not so simple to
discuss nor so easy to quantify as those brought to your attention previously.

Any active defense system such as the ABM must sit in readiness for
two or four or eight years and then fire at the precisely correct second following a warning time of only a few minutes. This warning time is so
short that systems designers usually attempt to eliminate human decision-
makers, even at low command levels, from the decision-making system.
Further, the precision needed for the firing time is so fine that machines
must be used to choose the precise instant of firing no matter how the decision to fire is made.

In the case of offensive missiles the situation is different in an essential
way: although maintaining readiness throughout a long, indefinite period
is necessary, the moment of firing is not so precisely controlled and hence
human decision-makers, including even those at high levels, can be per-

mitted to play a part in the decision-making process. Thus the trigger of any ABM, unlike the trigger of ICBMs and Polarises, must be continuously sensitive and ready, in short a "hair" trigger for indefinitely long periods of time. On the other hand, it is obvious that we cannot afford to have an ABM fire by mistake or in response to a false alarm, and indeed the army has recently gone to some pains to assure residents of areas near proposed Sentinel sites that it has imposed design requirements which will insure against the accidental launching of the missile and the subsequent detonation of the nuclear warhead it carries. These two requirements, a "hair" trigger so that it can cope with a surprise attack and a "stiff" trigger so that it will never go off accidentally, are, I believe, contradictory requirements.

This problem exists only in the real world and not on the test range; on the test range there need be no such concern about accidental misfires, the interceptions do not involve the use of nuclear weapons and the day, if not the second, of the mock attack is known. Another essential (but again difficult to quantify) difference between the real world and the test range lies in the fact that the deployed defensive equipment will, normally, never have been fully exercised and even the supposedly identical test-range equipment will never have been tested against the precise target or targets that the deployed equipment would ultimately have to face.

In the case of other defense systems which have worked after a fashion, practice using the actual deployed equipment against real targets has been possible and has been a major element in increasing their effectiveness. Thus, the Soviet SAMs (surface-to-air missiles) in North Vietnam work as well as they do because both the equipment designers and the operating crews have had plenty of opportunities to practice against U.S. targets equipped with real counter-measures and employing real tactics.

Capability Question

For these and similar reasons, as well as because of technical problems, I continue to have the gravest doubts as to the capability of any ABM system I have heard of, whether or not the problem has been defined into being "easy" and whether or not it "works" on a test range. I am not talking about some percentage failure inherent in the mathematical distribution of miss distances, nor statistically predictable failures in system components, but rather about catastrophic failure in which, at the moment of truth, either nothing happens at all, or all interceptions fail.

I should like now to turn from technical matters to political matters concerning the relationship between the ABM and arms control policies and possibilities. It is frequently said that the ABM, or at least some versions of it, does not have serious arms control implications, the reasons advanced having to do with its intrinsically defensive character. In my opinion such a belief is based on an error which may be called the "Fallacy of the Last Move." It is indeed true, in some cases, that if the last move that was ever made in the arms race were that of deploying an ABM system, then deploying the ABM would by definition not have any arms race implications. But in the real world, in which there currently is constant change in both the technology and the deployed numbers of all kinds of strategic systems, ABMs do have disarmament implications. In support of this notion, let me turn to a relevant bit of real recent history.

At the beginning of the Sixties, we began to hear about a possible Soviet ABM and we became concerned about its possible effects on our ICBM and Polaris systems. It was then that we began to consider seriously various penetration aid ideas among which was the notion of placing more than one warhead on a single offensive missile. This original idea has since grown in complexity, as these things do, and has resulted in the MIRV (multiple independent re-entry vehicle) concept.

There are now additional justifications for MIRV besides penetration, but that is how it all started. As others have pointed out, the MIRV concept is a very important element in the arms race, and potentially seriously destabilizing. In fact, the possibility of a Soviet MIRV is used as one of the main arguments in support of the idea of hard-point defense, and thus we have come one full turn around the arms race spiral. But no one in 1960-61 thought through the potential destabilizing effects of multiple warheads, and certainly no one did, or even could have, predicted that the inexorable logic of the arms race would carry us directly from Soviet talk in 1960 about defending Moscow against missiles, to a requirement for hard-point defense of *offensive* missile sites in the United States in 1969. Similarly no one today can outline in detail what kind of a chain-reaction a Sentinel or a hard-point defense system would lead to. But we all know of the propensity of scientists and engineers to respond to technical challenge with further technical complexity and we have seen the willingness of both sides to pay for the supposed technical solutions at almost any cost.

Implications for Arms Control

Thus, although I cannot be sure of the mechanism, I believe that either hard-point defense or Sentinel would produce further acceleration of the arms race. It is possible that the deployment of these ABMs would lead to greater numbers of deployed offensive warheads on both sides. We may expect deployment of these ABMs would lead to the persistent query, "But how do you know it *really* works?" And the pressures now applied against the current Partial Test Ban Treaty would be multiplied. It is certain that deployment of these ABMs would lead to more steps in that awesome direction of placing greater reliance on automatic devices for making that ultimate decision as to whether or not doomsday had arrived.

It thus appears that as a specific part of a well thought out and well defined arms control agreement, deployment of hard-point defense might play a positive role, but otherwise it would be just one more step away from national security.

Finally, perhaps the worst arms control implication of the ABM is the possibility that the people and the Congress would be deceived into believing that at long last we are on the track of a technical solution to the dilemma of the steady decrease in our national security which has accompanied the steady increase in our military power over the last two decades. Such a false hope is extremely dangerous if it diverts any of us from searching for a solution in the only place it may be found: in a political search for peace combined with arms control and disarmament measures.

In Summary

1. Because of certain intrinsic disadvantages of the defense, and because of certain fundamental design problems, I doubt the capability of either the Sentinel System or the hard-point defense ABM to accomplish its task, whether or not it ultimately "works" on a test range.

2. I believe the deployment of any ABM would in the long run almost always result in further acceleration of the arms race. An exception would be in the case of the deployment of an ABM as a carefully integrated part of a major move in the direction of arms control and disarmament.

3. One result of the arms race is that, as our military power increases, our national security decreases. I believe this basic situation would not be improved by deployment of any ABM.

4. Another result of the arms race is that, due to the ever-increasing complexity of both offensive and defensive systems, the power to make certain life-and-death decisions is inexorably passing from statesmen and politicians to more narrowly focused technicians, and from human beings to machines. An ABM deployment would speed up this process.

A Personal View
of the Arms Race

My entire professional life has been dominated by the nuclear arms race.

I entered college at the University of Rochester in September 1939, two weeks after the Germans initiated World War II by invading Poland. During my sophomore and junior years, my physics instructors began to disappear one by one into various secret war programs. Immediately after my own hurried up graduation, I too left and joined the group at Berkeley which was engaged in the separation of uranium isotopes. After those very isotopes were exploded over Hiroshima in 1945 in history's first military application of nuclear energy, I returned to a career in pure science, or so I thought at the time. Four years later in 1949 I received my Ph.D. degree, and only months after that, the Russians exploded their first atomic bomb. Soon after, I became deeply involved in the acceleration of the U.S. weapons program that resulted from our reaction to the first Soviet tests. In the years that followed I worked in a number of high administrative positions, sometimes deep within the arms race, sometimes outside, but even then always well within its shadow. Finally, I decided to drop that kind of activity and to try my hand once again at physics. To make a long story short, what happened was that in the Spring of 1969 I found myself once again completely immersed in the arms race, this time trying to do what I could to oppose the deployment of the ABM. I oppose that action because I believe it would be a grave mistake for a number of reasons, the most important being that it would seriously accelerate the arms race just at a time when there appears to be promise of getting it under control.

Let me briefly, and somewhat superficially review first, the facts of the arms race, second, what little has been done in the way of arms control, third, why it's so hard to do anything, and last why the present is such a critical time in the efforts to achieve some important degree of arms control and disarmament.

During the years immediately after World War II, the U.S. continued to produce missile material in the plants that had been set up or started during the war, and continued to fabricate this material into bombs of more or less the same design and characteristics. By about 1950, we had accumulated a stockpile of some hundreds of bombs, each having an explosive yield equivalent to a few tens of kilotons of TNT. The total yield of this stockpile, then, contained the equivalent of some few millions of tons (or megatons, as we say) of chemical explosives. This was a few times the total explosive energy of all the bombs dropped on Germany during World War II. Allowing for various detailed differences in the effects produced by explosions of different sizes, a stockpile of this size might be expected to cause about the same level of casualties and damage that occurred in Germany during World War II, except that, whereas that war spread these out over more than five years, they could now be caused to happen in a single day.

We might have gone on slowly accumulating A-bombs except for four closely spaced events which raised new anxieties and generated new possibilities. These were the first Russian A-test in 1949, the Korean War which followed soon thereafter, the successful U.S. thermonuclear, or "H-bomb," test in 1952, and lastly, only nine months later, the first Soviet thermonuclear test in 1953. These new thermonuclear explosives were, roughly speaking, one thousand times as powerful as the A-bombs which preceded them. Also during the early fifties, and on top of that increase in unit size, we accelerated our unit production rates. The result was that by about 1960, we had thousands of bombs, many of which had tens of megatons of explosive energy each. The net result was that we could then bring about the equivalent of tens of thousands of World War II's in a single day. We had by that time reached a level of explosive power which some writers have characterized by the word "overkill," an understatement in my opinion. We had achieved a kind of super saturated state of affairs in that the stockpile, if most of it were used, could destroy all the major population centers of our potential enemies many times over, killing nearly all of the urban population and much of the rural population in the process. The total area which could be bathed in lethal fallout, if the explosives were appropriately distributed, would be many millions of square miles, a situation which led to the kind of doomsday descriptions contained in such books as *On the Beach* by Nevil Shute.

After 1960, nuclear weapons development and stockpiling continued, but such efforts were largely directed to improvements and changes in qualitative features, such as adapting these explosives to different environments, including more hostile ones, adapting them to different types of delivery systems, and improving factors such as yield to weight ratios.

Meanwhile great changes in the mode of delivery of weapons were taking place. During and after World War II, in fact until 1960, the strategic delivery of nuclear weapons was accomplished solely by aircraft. Some rocket applications pre date 1960, but these involved short range air to air and surface to air defensive systems. The development programs which led to today's ICBM's got underway in a big way in 1953, and deployment of these followed the usual seven years later. In 1953, the technologies of rocket propulsion, thermonuclear explosives, and guidance and control reached a point such that it was possible to begin the development of intercontinental rockets which could do more or less the same overall job as long range aircraft, but do it an order of magnitude faster, and with an even greater certainty of penetrating any real or hypothetical defenses. At about the same time, we became aware of similar, perhaps more advanced, programs going on in the Soviet Union, and as a result of the conjunction of these two events, we began those huge crash programs in missile development which dominated the technological scene in the late fifties and early sixties. We simultaneously initiated the Air Force's Thor, Atlas and Titan missile programs and the Army's Jupiter missile program, and soon after we phased in the Navy's Polaris system and the Air Force's Minute Man. Serious work on the development of an anti-missile missile also was started in the mid-fifties by the Army. That program, called Nike Zeus at first, was a natural outgrowth of the Army's earlier anti-aircraft rocket programs: the Nike Ajax, and the Nike Hercules. As we now know, and insofar as deployment is concerned, we over reacted to what we took to be the Soviet threat. Neglecting the first few rockets which had no real significance in the strategic balance, we deployed many more rockets much faster than they did. We did so for a number of reasons: 1) the politically motivated "missile gap" charges of the political "outs" during the 1960 Presidential campaign, 2) the genuine fears of the political "ins" promoted by "worst case analysis" (which I will come to again below) and 3) the very great capacity of U.S. industry to react to a challenge of this sort when fueled with sufficient money. By the mid-sixties, and in addition to our long-range aircraft, both our ICBM force and our sub launched missile force separately contained many times the amount of explosive power needed for deterrence. Then, having reached a level of super-saturation in this area also, we leveled off our deployments in terms of numbers. However, just as in the case of nuclear weapons, we continued with development programs whose purposes were to improve qualitative features, such as the ratio of payload to gross rocket weight, accuracy, range, defense penetrability, and most important for its future significance, arranging things so that a number of separately targetable warheads could be launched by a single rocket booster. The name for this last technique is MIRV, an acronym standing for *M*ultiple *I*ndependently-targetable *R*eentry *V*ehicles.

Throughout this whole period beginning with the end of World War II, the Soviet Union, while following the same general path as the U.S., has always been behind in both quality and quantity of deployed weapons in an overall sense. There were of course individual examples of weapons systems where they were ahead for a time, but overall and as precisely as one can apply a numerical measure to these things, they were behind us by, typically, three years during the fifties and early sixties. Their deployment program, therefore, continued on after ours slowed down and leveled off. By 1969 they are just now catching up with us in numbers of ICBM's and we anticipated that they would catch up with us in sublaunched missiles some time in the next few years. Some people have seen this continuing increasing deployment of Soviet weapons after our own deployment had leveled off as something necessarily sinister, as proving that they meant to be able to do us in as soon as they could. No one can say for sure what their most private intentions are, but a simpler explanation of their continuing deployment is simply that they finally reacted to being so far behind for so long. Of course, they may very well have over-reacted just as we virtually always do when we get merely the first evidence that we may be in danger of falling behind.

Let me now leave the arms race, and turn to the other side of the coin, Arms Control and Disarmament. Sensitive and perceptive people everywhere, beginning immediately after the Hiroshima explosion, realized that the nuclear arms race posed a new kind and degree of danger to the human race. Even when only a few bombs in only the ten-kiloton range existed, and when these were in the hands of only a single nuclear power, some people saw that the day would come when the number of such weapons, the power of each individual weapon, and the proliferation of control over these weapons, would reach a point where man could destroy not just a few individual cities, but civilization itself, and ultimately perhaps, even the entire human race. Out of this concern, came the Baruch Plan and later, Eisenhower's Atoms for Peace Plan. On the Soviet side, and especially after they had successfully tested their own atomic weapons, proposals for slowing or stopping the nuclear arms race were also generated. The UN General Assembly tackled the problem, and set up special committees and passed resolutions on the subject. A variety of difficulties were encountered, but three particular stumbling blocks to the establishment of formal arms control treaties soon became evident. First, and generally speaking, the Soviet proposals and the UN resolutions involved agreements which would stop something right now, and leave the working out of policing and enforcement measures for later, whereas the U.S. put the emphasis on working out policing methods first to be followed by stopping whatever it was later. A second but closely related stumbling block consisted in the U.S. normally insisting on fairly large numbers of fairly extensive "on site inspections," whereas the Soviet Union nearly always rejected out of hand the idea of any at all. And thirdly, the fact that the U.S. was so far ahead in everything also proved to be a stumbling block in most instances: reducing arms or other stockpiles by equal amounts would always have resulted in making our advantage even greater, and the Soviets couldn't buy that, on the other hand reducing to equal amounts always involved greater sacrifice on the part of the U.S., and that was never politically palatable here.

Even so, some progress was made. Measured by the tenacious efforts of those few souls who labored in this particular vineyard, and considering the obstacles, the progress has been remarkable; compared to the great crescendo of the arms race itself, progress has hardly been noticeable. Out of these efforts have come five arms control treaties which involve both the U.S. and the U.S.S.R. The first of these was the Antarctic Treaty, signed in 1959. It forbade the militarization of Antarctica and prohibited nuclear testing there. Because it was so remote and no real vital interests were involved, it was possible to include an inspection clause in the treaty. The U.S. has actually exercised this right to inspect Soviet scientific bases there and thus has assured itself that these are not somehow secretly being converted into military bases. Following this came the Limited Test Ban Treaty in 1963, the Latin American Denuclearization Treaty in 1967, and also in 1967 the Outer Space Treaty which bars nuclear weapons testing and deployment in space or on the moon or other celestial bodies, and in fact, prohibits military bases of any kind on such celestial bodies. The most recent treaty is the Non Proliferation Treaty, opened for signing last July in Washington, London, and Moscow.

It is interesting to examine the Limited Test Ban Treaty more closely to see, by example, what some of the key problems have been and how they were resolved. Like all treaties in this area, it can be traced back to the Baruch Plan and to some of the earliest UN resolutions. However, perhaps the first specific government action, as distinguished from proposals, was taken by the Soviet Union, when in March 1958, the Supreme Soviet passed a decree abolishing Nuclear Testing in the U.S.S.R. Khrushchev reported this to Eisenhower a few days later and asked the U.S. to do the same. We already had a test series underway, and furthermore, we always had insisted on working out methods for policing a ban first, and so we did not comply. The Soviets complained, and said if we didn't stop, they would resume, and in fact they did so only five months later. Meanwhile, plans and hopes to hold an international meeting of "experts" in matters of nuclear development, testing, and test detection finally bore fruit that summer in Geneva. These meetings produced enough agreement so as to lead directly to the possibility of higher level political discussions later that fall. Secretary of State Dulles then offered, on behalf of the U.S., a one year moratorium on tests to begin when such political talks began, and October 31, 1958 was suggested as the starting date for them. Both President Harold Brown and Provost Robert Bacher were very deeply involved in both the Meeting of Experts and the subsequent Political Meetings. The U.S. did finish its test series before that date, and the Soviet Fall series ended on November 3. Serious talks did begin, and no further nuclear explosions took place. The most serious difficulties in these talks were associated with problems of policing a ban on underground testing and the purported need for on-site inspection of whatever suspicious events a seismic detection system might turn up. Finally, after talks of this sort had dragged on for a year, President Eisenhower announced that the U.S., having offered only a one year moratorium in the first place, was no longer obligated to refrain from testing, but would not resume without advance notice. The next day, Chairman Khrushchev, apparently in response, announced the Soviets would not resume testing unless the West did so first. These two statements were made in the closing days of 1959. In 1960, the French began nuclear tests in the Sahara, and the Soviets promptly labelled this as testing by the West, and charged that connivance by the U.S. and the United Kingdom was involved. One should recall here that both the Eastern and Western alliances still had at least the outward appearance of being monolithic at that time. Also, that same summer, the U-2 was shot down over Siberia, and the Paris summit conference between Eisenhower and Khrushchev collapsed. Still, neither the United States nor the Soviet Union tested any nuclear weapons. In 1961, the French continued to test, and Khrushchev commented to a U.S. emissary that his "scientists and military" were pressing for the test of a 100 MT bomb. I know from my own participation in government affairs at that time, that very high pressures in support of a resumption of testing by the U.S. were also building up in this country, and charges that the Soviets "must be cheating" were rife, though I personally never found the evidence at all convincing.

Finally, in the fall of 1961, the Soviets began a new series of nuclear tests in the atmosphere. Two weeks later the U.S. began a series of underground tests, and six months later both the U.S. and the U.K were testing in the atmosphere. Fallout began to fall again in larger amounts than ever before, passions became aroused all over the world, the UN General Assembly pressed the nuclear powers to do something, and further conferences of experts and statesmen were held. Finally, on July 1, 1963, five years after the first serious attempt at a test moratorium, the Limited Test Ban Treaty was signed. This treaty was "limited" in that it allowed underground testing while prohibiting testing in the atmosphere, underwater, and in outer space. The difficult problems of detecting, identifying, verifying and policing underground tests had been "solved" by bypassing them. It was, or course, the hope and intention of most of those involved that this part of the problem would be intensively attacked later, but precious little progress has been made on this matter since that time. In my opinion, exaggerated fears of possible cheating compounded by exaggerated ideas of what might be learned in one or two clandestine tests of small devices combined with exaggerated fears of what new technological breakthroughs might be made with such knowledge and compounded by further exaggerated ideas of what the political significance of such hypothetical new technology might be, have dominated the scene and hamstrung our efforts in this area. The Soviets have also had their hang-ups, but I cannot identify them so clearly.

All of this is another example of "worst case analysis," or what more commonly is called "erring on the side of military safety." These two phrases are the slogans or the mottos of the arms race. They can be and have been used to justify acceleration of our arms developments and deployments and to justify foot dragging in arms control. That way of looking at things makes it impossible ever to achieve a situation in the arms race which both sides would recognize as parity. In the face of the residual uncertainties which always exist about what an opponent is doing or plans to do, "worst case analysis" by each side inevitably leads to a gross exaggeration of both the capabilities and intent of the other. In the inner technical and planning circles where such analyses are made, there is some understanding of how improbable these worst plausible cases really are. However, as the analyses filter up through the various ever more political layers to the statesmen on top this distinction disappears and the worst plausible case comes to be taken as the probable case. When that happens bilateral arms control measures, as well as any unilateral steps in the direction of moderation, are blocked by an almost impenetrable barrier of fear and suspicion. "Erring on the side of military safety" sounds like prudent advice, but as actually used, it works out to be a justification for paralyzing any attempts at arms control.

Another kind of problem is illustrated by some of the discussions preceding the Non Proliferation Treaty (or NPT), which is an attempt by the U.S., U.K, and U.S.S.R. to limit the spread of nuclear weapons beyond the five nations now possessing them. This problem has to do with what is called Project Plowshare, which is a set of ideas about how to employ nuclear explosives for peaceful purposes. They are mostly, except for scale, analogous to the uses of chemical explosives: they range from cracking mineral rich rocks to preparing them for subsequent chemical leaching to blasting out harbors and canals. So far none of the really important ideas, in my view, concern anything which cannot be accomplished by conventional means, though in some cases it is thought to be cheaper in the long run to do it atomically. (I have serious doubts as to whether the nuclear excavation type projects are in fact as cheap as claimed. In the cases I've seen in detail, it seems that the costs associated with the evacuation of populations or living with radioactivity have been much underestimated.) It is obviously difficult to accommodate such a program in any straightforward way into either the Limited Test Ban or the Non Proliferation Treaty. Without going into details, one reason for this is that while the ends of such explosions are very different from those involved in military uses, the means are basically the same. Yet, even though after about fifteen years of work, nothing which would be of real or social or political importance and which cannot be done by other means has yet been proposed, the possibility that it might be has been a great drag on all nuclear control agreements. This is despite the fact that the future of civilization and conceivably even mankind itself is at stake on the one hand, whereas on the other there seems to be little more than the possibility of saving some money. I regard this as an extremely poor ordering of priorities, yet not only the U.S. and to a less degree the Soviet Union have dragged their feet on this issue, but other countries, notably India and Brazil have used these Plowshare possibilities as an excuse for causing delays in arms control agreements running into years.

Why so little? Why so slow? We've just gone over some example: fear of what the other side might be up to, fear of what they might gain by cheating, lack of complete confidence in the methods (unilateral or even bilateral) proposed for policing agreements, arguments over which comes first; the agreement to stop something followed by the working out of policing methods or vice versa, arguments over the number of on-site inspections and over how and by whom they ought to be carried out. This long list of worries reminds me of a remark said to have been made by J. Robert Oppenheimer in 1958, and published in both Izvestia and the Congressional Record in 1969: "The anxiety of the public must become sufficiently deep so that the idea of disarmament can take hold of the masses." The generalized anxiety of the public at present is clearly still insufficient to overcome the more detailed and better articulated worries of those who oppose every real particular attempt at disarmament while they proclaim their support of the general idea.

I remarked earlier that the present seems to be a particularly propitious time for achieving some new and stronger steps in the way of arms control and disarmament agreements. Let me now go further and add that I believe we are approaching a deadline in this matter, and if we don't successfully seize this opportunity quickly, it will pass and another, like it, will not come along for a long time.

I think the present period is propitious because the strategic arms situation is currently characterized by two factors both of which are essential to achieving meaningful strategic arms limitations; I think the present period is critical because one of these factors is transitory. The first of these factors which characterize the present situation is parity. Here I mean parity not only in the sense that each side has (or is about to have) more or less equal numbers of weapons, but in the more important sense that each side has easily enough weapons to deter the other from attempting a first strike, and neither side has anywhere near enough weapons to strike such a devastating first blow that it could hope to survive even the residual retaliation of the other.

The second of these factors is the relatively high degree of confidence each side now has in the accuracy of its assessment of the other side's capabilities. The weapons systems now deployed (ICBM's in silos, bombers on military airfields, and missile bearing submarines) have physical characteristics such that each side can with considerable confidence and by unilateral means estimate both the numbers and the effectiveness of the other's weapons with sufficient accuracy so that each party can be reasonably certain about what could happen in various possible situations. I may note here in support of this claim that in all the recent ABM debates no one on either side of the argument seriously questioned the statements about how many weapons of the various types there really were currently deployed, and no one took issue with the estimates about their explosive power. The only major arguments about characteristics concerned future possibilities in the area of accuracy, and that matter is relatively less important as long as each side has deterrence of war as its goal.

It is a pending change in this second circumstance, that is, in the relative confidence in the information about the numbers and the effectiveness of the weapons the other side possesses, which threatens to end the present period of parity and stability during which negotiations seem to be so promising. The principle devices on the horizon which threaten to do this are the MIRV and the ABM (Recall MIRV stands for Multiple Independently-targetable Reentry Vehicles). Numbers of warheads per missile of more than ten have been talked about. Since it is much harder to know precisely what a rocket may have as a warhead or warheads, than it is to know simply that the large rocket exists and is in place, the introduction of MIRV into the presently deployed weapons systems will greatly decrease the confidence with which each side can estimate what the other can do. ABM compounds this problem. ABM is an exceedingly complex system, and if the recent debates have indicated anything, they have shown that its effectiveness may be virtually anywhere between a flat zero percent and nearly 100 percent. Thus, neither side can be at all sure of the performance of either its own or the other side's system. Even more than MIRV therefore, the ABM would, if deployed extensively, introduce very great uncertainty into any future attempts to estimate what could happen if the button were to be pushed. In the view of most of us who have lived through past attempts at arms control negotiations, such gross uncertainties and the worries that would flow from them even before being massaged by "worst case analysts" would severely inhibit if not entirely prevent further steps toward arms control. They would instead strongly induce further moves along the arms race spiral. It is for this reason that I judge the period just ahead to be so unusually critical.

Military Technology and National Security

I n the spring of 1969 public hearings in the Senate and the House of Representatives on anti-ballistic-missile (ABM) systems provided an unprecedented opportunity to expose to the people of this country and the world the inner workings of one of the dominant features of our time: the strategic arms race. Testimony was given by a wide range of witnesses concerning the development and deployment of all kinds of offensive and defensive nuclear weapons; particular attention was paid to the interaction between decisions in these matters and the dynamics of the arms race as a whole.

In my view the ABM issue is only a detail in a much larger problem: the feasibility of a purely technological approach to national security. What makes the ABM debate so important is that for the first time it has been possible to discuss a major aspect of this larger problem entirely in public. The reason for this is that nearly all the relevant facts about the proposed ABM systems either are already declassified or can easily be deduced from logical concepts that have never been classified. Thus it has been possible to consider in a particular case such questions as the following:

1. To what extent is the increasing complexity of modern weapons systems and the need for instant response causing strategic decision-making authority to pass from high political levels to low military command levels, and from human beings to machines?

2. To what extent is the factor of secrecy combined with complexity leading to a steadily increasing dominance of military-oriented technicians in some vital areas of decision-making?

3. To what extent do increasing numbers of weapons and increasing complexity—in and of themselves—complicate and accelerate the arms race?

My own conclusion is that the ABM issue constitutes a particularly clear example of the futility of searching for technical solutions to what is essentially a political problem, namely the problem of national security. In support of this conclusion I propose to review the recent history of the strategic arms race, to evaluate what the recent hearings and other public discussions have revealed about its present status and future prospects, and then to suggest what might be done now to deal with the problem of national security in a more rational manner.

The strategic arms race began in the early 1950's, when it became evident that the state of the art in nuclear weaponry, rocket propulsion and missile guidance and control had reached the point in the U.S. where a strategically useful intercontinental ballistic missile (ICBM) could be built. At about the same time the fact that a major long-range-missile development program was in progress in the U.S.S.R. was confirmed. As a result of the confluence of these two events the tremendous U.S. long-range-missile program, which dominated the technological scene for more than a decade, was undertaken. The Air Force's Thor, Atlas and Titan programs and the Army's Jupiter program were started almost simultaneously; the Navy's Polaris program and the Air Force's Minuteman program were phased in just a few years later.

More or less at the same time the Army, which had had the responsibility for ground-based air defense (including the Nike Ajax and Nike Hercules surface-to-air missiles, or SAM's), began to study the problem of how to intercept ICBM's, and soon afterward initiated the Nike Zeus program. This program was a straightforward attempt to use existing technology in the design of a nuclear-armed rocket for the purpose of intercepting an uncomplicated incoming warhead. The Air Force proposed more exotic solutions to the missile-defense problem, but these were subsequently absorbed into the Defender Program of the Department of Defense's Advanced Research Projects Agency (ARPA). The Defender Program included the study of designs more advanced than Nike Zeus, and it also incorporated a program of down-range measurements designed to find out what did in fact go on during the terminal phases of missile flight.

By 1960 indications that the Russians were taking the ABM prospect seriously, in addition to progress in our own Nike Zeus program, stimulated our offensive-missile designers into seriously studying the problems of how to penetrate missile defenses. Very quickly a host of "penetration aid" concepts came to light: light and heavy decoys, including balloons, tank fragments and objects resembling children's jacks; electronic countermeasures, including radar-reflecting clouds of the small wires called chaff; radar blackout by means of high-altitude nuclear explosions; tactics such as barrage, local exhaustion and "rollback" of the defense, and, most important insofar as the then unforeseen consequences were concerned, the notion of putting more than one warhead on one launch vehicle. At first this notion simply involved a "shotgun" technique, good only against large-area targets (cities), but it soon developed into what we now call MIRV's (multiple independently targeted reentry vehicles), which can in principle (and soon in practice) be used against smaller, harder targets such as missile silos, radars and command centers.

This avalanche of concepts forced the ABM designers to go back to the drawing board, and as a result the Nike-X concept was born in 1962. The Nike-X designers attempted to make use of more sophisticated and up-to-date technology in the design of a system that they hoped might be able to cope with a large, sophisticated attack. All through the mid-1960's a vigorous battle of defensive concepts and designs versus offensive concepts and designs took place. This battle was waged partly on the Pacific Missile Range but mostly on paper and in committee meetings.

This intellectual battle culminated in a meeting that took place in the White House in January, 1967. In addition to President Johnson, Secretary of Defense Robert S. McNamara and the Joint Chiefs of Staff there were present all past and current Special Assistants to the President for Science and Technology (James R. Killian, Jr., George B. Kistiakowsky, Jerome B. Wiesner and Donald F. Hornig) and all past and current Directors of Defense Research and Engineering (Harold Brown, John S. Foster, Jr., and myself). We were asked that simple kind of question which must be answered after all the complicated ifs, ands and buts have been discussed: "Will it work?" The answer was no, and there was no dissent from that answer. The context, of course, was the Russian threat as it was then interpreted and forecast, and the current and projected state of our own ABM technology.

Later that year Secretary McNamara gave his famous San Francisco speech in which he reiterated his belief that we could not build an ABM system capable of protecting us from destruction in the event of a Russian attack. For the first time, however, he stated that he did believe we could build an ABM system able to cope with a hypothetical Chinese missile attack, which by definition would be "light" and uncomplicated. In recommending that we go ahead with a program to build what came to be known as the Sentinel system, he said that "there are *marginal* grounds for concluding that a light deployment of U.S. ABM's against this possibility is prudent." A few sentences later, however, he warned: "The danger in deploying this relatively light and reliable Chinese-oriented ABM system is going to be that pressures will develop to expand it into a heavy

Soviet-oriented ABM system." The record makes it clear that he was quite right in this prediction.

Meanwhile the U.S.S.R. was going ahead with its own ABM program. The Russian program proceeded by fits and starts, and our understanding of it was, as might be supposed in such a situation, even more erratic. It is now generally agreed that the only ABM system the Russians have deployed is an area defense around Moscow much like our old Nike Zeus system. It appears to have virtually no capability against our offense, and it has been, as we shall see below, extremely counterproductive insofar as its goal of defending Moscow is concerned.

Development and deployment of offensive-weapons systems on both sides progressed rapidly during the 1960's but rather than discuss these historically I shall go directly to the picture that the Administration has given of the present status and future projection of such forces.

Data presented in 1969 by the Department of Defense show that the U.S. and the U.S.S.R. were about even in numbers of intercontinental missiles, and the U.S. was ahead in both long-range aircraft and submarines of the polaris type. The small Russian missiles were mostly what we call SS-11's, which were described in the hearings as being roughly the equivalent of our Minutemen. The large Russian missile was what we call the SS-9. Deputy Secretary of Defense David Packard characterized its capability as one 20-megaton warhead or three five-megaton warheads. Our own missiles are almost entirely the smaller Minutemen. There currently remain only 54 of the larger Titans in our strategic forces. Not covered in the table are "extras" such as the U.S.S.R.'s FOBS (fractional orbital bombardment system) and IRBM's (intermediate-range ballistic missiles), nor the U.S.'s bombardment aircraft deployed on carriers and overseas bases in Europe and elsewhere. There are, of course, many important details that do not come out clearly in such a simple tabular presentation; these include payload capacity, warhead yield, number of warheads per missile and, often the most important, warhead accuracy.

In the area of defensive systems designed to cope with the offensive systems outlined above, both the U.S. and the U.S.S.R. have defenses against bombers that would probably be adequate against a prolonged attack using chemical explosives (where 10 percent attrition is enough) and almost certainly inadequate against a nuclear attack (where 10 percent penetration is enough). In addition the U.S.S.R. has its ineffective ABM deployment around Moscow, usually estimated as consisting of fewer than 100 antimissile missiles.

What all these complicated details add up to can be expressed in a single word: parity. This is clearly not numerical equality in the number of warheads or in the number of megatons or in the total "throw weight"; in fact, given different design approaches on the two sides, simultaneous equality in these three figures is entirely impossible. It is, rather, parity with respect to strategic objectives; that is, in each case these forces are easily sufficient for deterrence and entirely insufficient for a successful preemptive strike. In the jargon of strategic studies either side would retain, after a massive "first strike" by the other, a sufficiently large "assured destruction capability" against the other in order to deter such a first strike from being made.

There is much argument about exactly what it takes in the way of "assured destruction capability" in order to deter, but even the most conservative strategic planners conclude that the threat of only a few hundred warheads exploding over population and industrial centers would be sufficient for the purpose. The large growing disparity between the number of warheads needed for the purpose and the number actually possessed by each side is what leads to the concept of "overkill." If present trends continue, in the future all or most missiles will be MIRVed, and so this overkill will be increased by perhaps another order of magnitude.

Here let me note that it is sometimes argued that there is a disparity in the present situation because Russian missile warheads are said to be bigger than U.S. warheads, both in weight and megatonnage; similarly, it is argued that MIRVing does not increase overkill because total yield is reduced in going from single to multiple warheads. This argument is based on the false notion that the individual MIRV warheads of the future will be "small" when measured against the purpose assigned to them. Against large, "soft" targets such as cities bombs *very much* smaller than those that could be used as components of MIRV's are (and in the case of Hiroshima were proved to be) entirely adequate for destroying the heart of a city and killing hundreds of thousands of people. Furthermore, in the case of small, "hard" targets such as missile silos, command posts and other military installations, having explosions bigger than those for which the "kill," or crater, radius slightly exceeds "circular error probably" (CEP) adds little to the probability of destroying such targets. Crater radius depends roughly on the cube root of the explosive power; consequently, if during the period when technology allows us to go from one to 10 warheads per missile it also allows us to improve accuracy by a little more than twofold, the "kill" per warhead will remain nearly the same in most cases, whereas the number of warheads increases tenfold.

In any case, it is fair to say that in spite of a number of such arguments about details, nearly everyone who testified at the ABM hearing agreed that the present situation is one in which each side possesses forces adequate to deter the other. In short, we now have parity in the only sense that ultimately counts.

Several forecasts have been made of what the strategic-weapons situation will be in the mid-1970's. In most respects here again there is quite general agreement. Part of the presentation by Deputy Secretary Packard to the Senate Foreign Relations Committee on March 26 were two graphs showing the trends in numbers of deployed offensive missiles beginning 1965 and extending to 1975. There is no serious debate about the basic features of these graphs. It is agreed by all that in the recent past the U.S. has been far ahead of the U.S.S.R. in all areas, and that the Russians began a rapid deployment program a few years ago that will bring them even with us in ICBM's quite soon and that, if extended ahead without any slowdown, would bring them even in submarine-launched ballistic missiles (SLBM's) sometime between 1971 and 1977.

One important factor that the Department of Defense omitted from its graphs is MIRV. Deployment plans for MIRV's have not been released by either the U.S. or the U.S.S.R., although various rough projections were made at the hearings about numbers of warheads per vehicle (three to 10), about accuracies (figures around half a mile were often mentioned, and it was implied that U.S. accuracies were better than Russian ones) and about development status (the U.S. was said to be ahead in developments in this field). A pair of charts emphasizing the impact of MIRV was prepared by the staff of the Senate Foreign Relations Committee.

One could argue with both of these sets of charts. For example, one might wonder why the Senate charts show so few warheads on the Russian Polaris-type submarine and why they show only three MIRV's on U.S. Minutemen; on the other hand, one might wonder whether the Department of Defense's projected buildup of the Russian Polaris fleet could be that fast, or whether one should count the older Russian missile submarines. Nonetheless, the general picture presented cannot be far wrong. Moreover, the central arguments pursued throughout the ABM hearings (in both the Senate Foreign Relations Committee hearings in March and the Senate Armed Services Committee hearing in April) were not primarily concerned with these numerical matters. Rather, they were concerned with (1) Secretary of Defense Melvin R. Laird's interpretation of these numbers insofar as Russian intentions were concerned, (2) the validity of the Safeguard ABM system as a response to the purported strategic problems of the 1970's and (3) the arms-race implications of Safeguard.

As for the matter of intentions, those favoring the ABM concept generally held that the only "rational" explanation of the Russians' recent SS-9 buildup, coupled with their multiple-warhead development program and

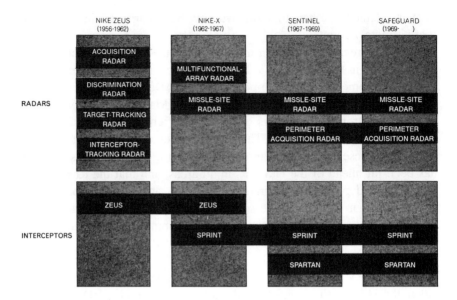

Figure 1. Evolution of U.S. ABM Systems is represented in this illustration, which is adapted from a chart introduced by Daniel J. Fink in his testimony before the Senate Foreign Relations Committee on March 6, 1969. In general the radar components of the successive designs have progressed from slow, mechanically steered, single-function radars to fast, electronically steered, multifunction radars. The slow Zeus ABM missile has been superseded by the short-range Sprint (for terminal defense) and the long-range Spartan (for area defense). The components of the Safeguard system are the same as those that were originally intended for the earlier Sentinal system.

the Moscow ABM system, was that they were aiming for a first-strike capability. One must admit that almost anything is conceivable as far as intentions are concerned, but there certainly are simpler, and it seems to me much more likely, explanations. The simplest of all is contained in Deputy Secretary Packard's chart. The most surprising feature of this chart is the fact that the Russians were evidently satisfied with being such a poor second for such a long time. This is made more puzzling by the fact that all during this period U.S. defense officials found it necessary to boast about how far ahead we were in order to be able to resist internal pressures for still greater expansion of our offensive forces.

Another possible reason, and one that I believe added to the other in the minds of the Russian planners, was that their strategists concluded in the mid-1960's that, whatever the top officials here might say, certain elements would eventually succeed in getting a large-scale ABM system

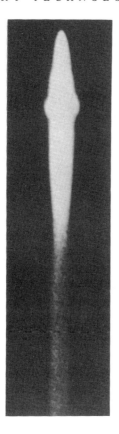

Figure 2. Sprint ABM missle was photographed during a test flight at the White Sands Missile Range in New Mexico. The photograph, which was released by the U.S. Army's Nike-X Project Office in March, 1966, shows the second stage of the missile heated to incandescense by friction with the atmosphere. Because of the extremely high speed of the Sprint, the missile's skin in places reaches temperatures hotter than inside its rocket motor. The bulges are created by the guidance fins at the rear of the second stage.

built, and the penetration-aid devices, including multiple warhead, would be needed to meet the challenge. Whether or not they were correct in this latter hypothetical analysis is still uncertain at this writing. Let us, however, pass on from this question of someone else's intentions and consider whether or not the Safeguard ABM system is a valid, rational and necessary response to the Russian deployments and developments outlined above.

To many of those who have written favorably about ABM defenses or who have testified in their favor before the Congressional committees, Safeguard is supported mainly as a prototype of something else: a "thick" defense of the U.S. against a massive Russian missile attack. This is clearly not at all the rationale for the Safeguard decision as presented by President Nixon in his press conference of March 14, nor is it implied as more than a dividend in the Defense secretaries' testimony. The President said that he wanted a system that would protect a part of our Minuteman force in order to increase the credibility of our deterrent, and that he had overruled moving in the direction of a massive city defense because "even

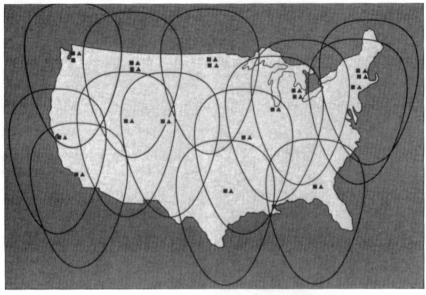

▲ SPARTAN SITE   ▲ SPRINT SITE   ■ MSR SITE   ■ PAR SITE

Figure 3. Sentinel system was decribed by the Johnson Administration as a "thin" ABM system designed to defend the U.S. against a hypothetical Chinese missile attack in the 1970's. The main defense was to be provided by long-range Spartan missiles. The Spartans would be deployed at about 14 locations in order to provide an area defense of the whole country. The range of each "farm" of Spartans is indicated by the egg-shaped area around it; for missiles attacking over the northern horizon the intercept range of the Spartan is elongated somewhat to the south. The Sentinel system would also include some short-range Sprint missiles, which were originally to be deployed to defend the five or six perimeter acquisition radars, or PAR's, which were to be deployed at five sites located across the northern part of the country. Missile-site radars, or MSR's, were to be deployed at every ABM site.

starting with a thin system and then going to a heavy system tends to be more provocative in terms of making credible a first-strike capability against the Soviet Union. I want no provocation which might deter arms talks." The top civilian defense officials give this same rationale, although they put a little more emphasis on the "prototype" and "growth potential" aspects of the system. For simplicity and clarity I shall focus on the Administration's proposal, as stated in open session by responsible officials.

From a technical point of view and as far as components are concerned, President Nixon's Safeguard system is very little different from President Johnson's Sentinel system. There are only minor changes in the

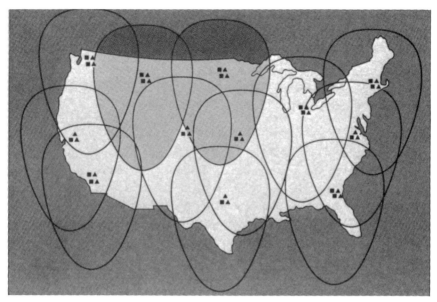

▲ SPARTAN SITE   ▲ SPRINT SITE   ■ MSR SITE   ■ PAR SITE

Figure 4. Safeguard system, President Nixon's proposed modification of the Sentinel scheme, uses essentially the same components in a slightly different array to accomplish an entirely different primary purpose: The defense of a part of our Minuteman force against a hypothetical surprise attack by the Russians. Phase I of Safeguard covers the construction of ABM sites at two Minuteman "fields": one near Malmstrom Air Force Base in Montana and the other near Grand Forks Air Force Base in North Dakota (shaded areas). The completed system would have a total of 12 sites, each with Sprint and Spartan coverage, located somewhat farther away from the cities. In addition two new PAR sites would be included in order to observe submarine-launched missiles coming from directions other than due north.

location of certain components (away from cities), and elements have been added to some of the radars so that they can now observe submarine-launched missiles coming from directions other than directly from the U.S.S.R. and China. As before, the system consists of a long-range interceptor carrying a large nuclear weapon (Spartan), a fast short-range interceptor carrying a small nuclear weapon (Sprint), two types of radar (perimeter acquisition radar, or PAR, and missile-site radar, or MSR), a computer for directing the battle, and a command and control system for integrating Safeguard with the national command. I shall not describe the equipment in detail at this point but pass on directly to what I believe can be concluded from the hearings and other public sources about each of the following four major questions: (1) Assuming that Safeguard could protect Minuteman, is it needed to protect our deterrent? (2) Assuming that safeguard "works," can it in fact safeguard Minuteman? (3) Will it work? (4) Anyway, what harm can it do?

First: Assuming that Safeguard could protect Minuteman, is it needed to protect our deterrent?

Perhaps the clearest explanation of why the answer to this first question is "no" was given by Wolfgang K.H. Panofsky before the Senate Armed Services Committee on April 22. He described how the deterrent consists of three main components: Polaris submarines, bombers and land-based ICBM's. Each of these components alone is capable of delivering far more warheads than is actually needed for deterrence, and each is currently defended against surprise destruction in a quite different way. ICBM's are in hard silos and are numerous. Polarises are hidden in the seas. Bombers can be placed on various levels of alert and can be dispersed.

Since the warning time in the case of an ICBM attack is generally taken as being about 30 minutes, the people who believe the deterrent may be in serious danger usually imagine that the bombers are attacked by missile submarines, and therefore have only a 15-minute warning. This is important because a 30-minute warning gives the bombers ample time to get off the ground. In that case, however, an attack on all three components cannot be made simultaneously; that is, if the attacking weapons are launched simultaneously, they cannot arrive simultaneously, and vice versa.

Thus it is incredible that all three of our deterrent systems could become vulnerable in the same period, and it is doubly incredible that we could not know that this would happen without sufficient notice so that we could do something about it. There is, therefore, no basis for a frantic reaction to the hypothetical Russian threat to Minuteman. Still, it is sensible and prudent to begin thinking about the problem, and so we turn to the other questions. We must consider these questions in the technological framework of the mid-1970's, and we shall do this now in the way defense officials currently seem to favor: by assuming that this is the best of all possible technological worlds, that everything works as intended and that direct extrapolations of current capabilities are valid.

Second: assuming that Safeguard "works," can it in fact safeguard Minuteman?

One good approach to this problem is the one used by George W. Rathjens in his testimony before the Senate Armed Services Committee on April 23. His analysis took as a basis of calculation the implication in Secretary Laird's testimony that the Minuteman force may become seriously imperiled in the mid-1970's. Rathjens then estimated how many SS-

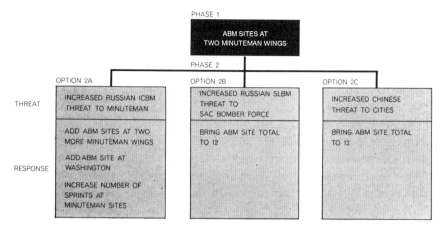

Figure 5. Safeguard Phase II provides for three different optional responses to various potential threats in the 1970's. A possible further addition would be sites in Alaska and Hawaii. This chart is also adapted from Deputy Secretary Packard's testimony on March 26.

9's would have to be deployed at that time in order to achieve this result. From this number, and the estimate of the current number of SS-9's deployed, he got a rate of deployment. He also had to make an assumption about how many Sprints and Spartans would be deployed at that time, and his estimates were based on the first phase of Safeguard deployment. These last numbers have not been released, but a range of reasonable values can be guessed from the cost estimates given. Assuming that the SS-9's would have four or five MIRV warheads each by that time, Rathjens found that by prolonging the SS-9 production program by a few months the Russians would be able to cope with Safeguard by simply exhausting it and would still have enough warheads left to imperil Minuteman, if that is indeed their intention.

The length of this short safe period does depend on the numbers used in the calculations, and they of course can be disputed to a degree. Thus if one assumes that it takes fewer Russian warheads to imperil Minuteman (it can't be less than one for one!), then the assumed deployment rate is lower and the safe period is lengthened; on the other hand, if one notes that the missile-site radars in our system are much softer than even today's silos, the first attacking warheads, fired directly at the radars, can be smaller and less accurate, so that a higher degree of MIRVing can be used for attacking these radars and a shorter safe period results. To go further, it was suggested that the accuracy/yield combination of the more numerous SS-11's might be sufficient for attacking the missile-site radars, and therefore, if the Russians were to elect such an option, there would be no safe period at all. In short, the most that Safeguard can do is either delay somewhat the date when Minuteman would be imperiled or cause the attacker to build up his forces at a somewhat higher rate if indeed imperiling Minuteman by a fixed date is his purpose.

In the more general case this problem is often discussed in budgetary terms, and the "cost-exchange ratio" between offense and defense is computed for a wide variety of specific types of weapon. Such calculations give a wide variety of results, and there is much argument about them. However, even using current offense designs (that is, without MIRV), such calculations usually strongly favor the offense. This exchange ratio varies almost linearly with the degree of MIRVing of the offensive missiles, and therefore it seems to me that in the ideal technological future we have taken as our context this exchange ratio will still more strongly favor the offense.

Third: Will it work? By this question I mean: Will operational units be able to intercept enemy warheads accompanied by enemy penetrations aids in an atmosphere of total astonishment and uncertainty? I do not mean: Will test equipment and test crews intercept U.S. warheads accompanied by U.S. penetrations aids in a contrived atmosphere? A positive answer to the latter question is a necessary condition for obtaining a positive answer to the former, but it is by no stretch of the imagination a sufficient condition.

This basic question has been attacked from two quite different angles: by examining historical analogies and by examining the technical elements of the problem in detail. I shall touch on both here. Design-oriented people who consider this a purely technical question emphasize the second approach. I believe the question is by no means a purely technical question, and I suggest that the historical-analogy approach is more promising, albeit much more difficult to use correctly.

False analogies are common in the argument. We find that some say: "You can't tell me that if we can put a man on the moon we can't build an ABM." Others say: "That's what Oppenheimer told us about the hydrogen bomb." These two statements contain the same basic error. They are examples of successes in a contest between technology and nature, whereas the ABM issue involves a contest between two technologies: offensive weapons and penetration aids versus defensive weapons and discrimination techniques. These analogies would be more pertinent if, in the first case, someone were to jerk the moon away just before the astronauts landed, or if, in the second case, nature were to keep changing the nuclear-reaction probabilities all during the development of the hydrogen bomb and once again after it was deployed.

Proper historical analogies should involve modern high-technology defense systems that have actually been installed and used in combat. If one examines the record of such systems, one finds that they do often produce some attrition of the offense, but not nearly enough to be of use against a nuclear attack. The most up-to-date example is provided by the Russian SAM's and other air-defense equipment deployed in North Vietnam. This system "works" after a fashion because both the equipment designers and the operating crews have had plenty of opportunities to practice against real U.S. targets equipped with real U.S. countermeasures and employing real U.S. tactics.

The best example of a U.S. system is somewhat older, but I believe it is still relevant. It is the SAGE system, a complex air-defense system designed in the early 1950's. All the components worked on the test range, but by 1960 we came to realize, even without combat testing, that SAGE could not really cope with the offense that was then coming into being. We thereupon greatly curtailed and modified our plans, although we did continue with some parts of the system. To quote from the recent report on the ABM decision prepared by Wiesner, Abram Chayes and others: "Still, after fifteen years, and the expenditure of more than $20 billion, it is generally conceded that we do not have a significant capability to defend ourselves against a well-planned air attack. The Soviet Union, after even greater effort, has probably not done much better."

So much for analogies; let us turn to the Safeguard system itself. Doubts about its being able to work were raised during the public hearings on a variety of grounds, some of which are as follows:

First, and perhaps foremost, there is the remarkable fact that the new Safeguard system and the old Sentinel system use virtually the same hardware deployed in a very similar manner, and yet they have entirely different primary purposes. Sentinel had as its purpose defending large soft targets against the so-called Chinese threat. The Chinese threat by definition involved virtually no sophisticated penetration aids and no possibilities of exhausting the defense; thus were "solved" two of the most difficult problems that had eliminated Nike Zeus and Nike-X.

Safeguard has as its primary purpose defending a part of the Minuteman force against a Russian attack. It is not credible that a Russian attack against the part of the Minuteman force so defended would be other than massive and sophisticated, so that we are virtually right back to trying to do what in 1967 we said we could not do, and we are trying to do it with no real change in the missiles or the radars. It is true that defending hard points is to a degree easier than defending cities because interception can be accomplished later and at lower altitudes, thus giving discrimination techniques more time to work. Moreover, only those objects headed for specific small areas must be intercepted. These factors do make the problem somewhat easier, but they do not ensure its solution, and plenty of room for doubt remains.

Second, there is the contest between penetration aids and discrimination techniques. The Russian physicist Andrei D. Sakharov, in his essay "Thoughts on Progress, Coexistence and Intellectual Freedom," put the issue this way: "Improvements in the resistance of warheads to shock waves and the radiation effects of neutron and X-ray exposure, the possibility of mass use of relatively light and inexpensive decoys that are virtually indistinguishable from warheads and exhaust the capabilities of an antimissile defense system, a perfection of tactics of massed and concentrated attacks, in time and space, that overstrain the defense detection centers, the use of orbital and fractional-orbital attacks, the use of active and passive jamming and other methods not disclosed in the press—all of this has created technical and economic obstacles to an effective missile defense that, at the present time, are virtually insurmountable."

I would add only MIRV to Sakharov's list. Pitted against this plethora of penetration aids are various observational methods designed to discriminate the real warheads. Some of the penetration devices obviously work only at high altitudes, but even these make it necessary for the final "sorting" to be delayed, and thus they still contribute to making the defense problem harder. Other devices can continue to confuse the defense even down to low altitudes. Some of the problems the offense presents to the defense can no doubt be solved (and have been solved) when considered separately and in isolation. That is, they can be solved for a time, until the offense designers react. One must have serious reservations, however, whether these problems can ever be solved for any long period in the complex combinations that even a modestly sophisticated attacker can present. Further, such a contest could result in a catastrophic failure of the system in which all or nearly all interceptions fail.

Third, there is the unquantifiable difference between the test range and the real world. The extraordinary efforts of the Air Force to test operationally deployed Minutemen show that it too regards this as an important problem. Moreover, the tests to date do seem to have revealed important weaknesses in the deployed forces. The problem has many aspects: the possible differences between test equipment and deployed equipment; the certain differences between the offensive warheads and penetration aids supplied by us as test targets and the corresponding equipment and tactics

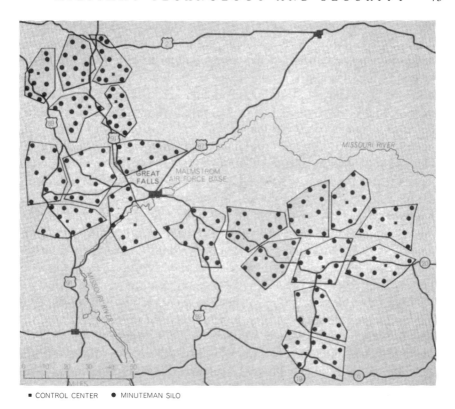

■ CONTROL CENTER ● MINUTEMAN SILO

Figure 6. Minuteman missiles Base in the vicinity of Malmstrom Air Force Base is shown on this map, which is based on information released by the Department of the Air Force. The minuteman missiles are grouped in 20 flights of 10 missiles each for a total of 200 missiles. Every flight has its own control center, each of which is capable of launching an entire squadron of 50 missiles.

the defense must ultimately be prepared to face; the differences between the installation crews at a test site and at a deployment site; the differences in attitudes and motivation between a test crew and an operational crew (even if it is composed of the same men); the differences between men and equipment that have recently been made ready and whom everyone is watching and men and equipment that have been standing ready for years during which nothing happened; the differences between the emotional atmosphere where everyone knows it is not "for real" and the emotional atmosphere where no one can believe what he has just been told. It may be that all that enormously complex equipment will be ready to work the very first time it must "for real," and it may be that all those thousands of human beings have performed all their interlocking assignments correctly, but I have very substantial doubts about it.

Fourth, there is the closely related "hair-trigger/stiff-trigger" contradiction. Any active defense system such as Safeguard must sit in readiness for two or four or eight years and then fire at precisely the correct second following a warning time of only minutes. Furthermore, the precision needed for the firing time is so fine that machines must be used to choose the exact instant of firing no matter how the decision to fire is made. In the case of offensive missiles the situation is different in an essential way: Although maintaining readiness throughout a long, indefinite period is necessary, the moment of firing is not so precisely controlled in general and hence human decision-makers, including even those at high levels, may readily be permitted to play a part in the decision-making process. Thus if we wish to be certain that the defense will respond under conditions of surprise, the trigger of the ABM system, unlike the triggers of ICBM's and Polarises, must be continuously sensitive and ready—in short, a hair trigger—for indefinitely long periods of time.

On the other hand, it is obvious that we cannot afford to have an ABM missile fire by mistake or in response to a false alarm. Indeed, the Army went to some pains to assure residents of areas near proposed Sentinel sites that it was imposing requirements to ensure against the accidental launching of the missile and the subsequent detonation of the nuclear warhead it carries. Moreover, Army officials have assured the public that no ABM missiles would ever be launched without the specific approval of "very high authorities."

These two requirements—a hair trigger so that the system can cope with a surprise attack and a stiff trigger so that it will never go off accidentally or without proper authorization—are, I believe, contradictory requirements. In saying this I am not expressing doubt about the stated intentions of the present Army leaders, and I strongly endorse the restrictions implied in their statements. I am saying, however, that if the system cannot be fired without approval of "the highest authorities," then the probability of its being fired under conditions of surprise is less than it would be otherwise. This probability depends to a degree on the highly classified technical details of the Command and Control System, but in the last analysis it depends more on the fact that "the highest authority" is a human being and therefore subject to all the failures and foibles pertaining thereto.

This brings us to our fourth principal question: Anyway, what harm can it do?

We have just found that the total deterrent is very probably not in peril, that the Safeguard system probably cannot safeguard Minuteman even if it "works," that there is, to say the least, considerable uncertainty whether or not it will "work." Nonetheless, if there were no harm in it, we might be prudent and follow the basic motto of the arms race: "Let us err on the side of military safety." There seem to be many answers to the question of what harm building an ABM system would do. First of all, such a system would cost large sums of money needed for nondefense purposes. Second, it would divert money and attention from what may be better military solutions to the strategic problems posed by the Administration. Third, it would intensify the arms race. All these considerations were discussed at the hearings; I shall comment here only on the third, the arms-race implications of the ABM decision.

It is often said that an ABM system is not an accelerating element in the arms race because it is intrinsically defensive. For example, during the hearings Senator Henry M. Jackson of Washington, surely one of the best-informed senators in this field, said essentially that, and he quoted Premier Kosygin as having said the same thing. I believe such a notion is in error and is based on what we may call "the fallacy of the last move." I believe that in the real world of constant change in both the technology and the deployed numbers of all kinds of strategic-weapons systems, ABM systems are accelerating elements in the arms race. In support of this view let us recall one of the features of the history recited at the start of this article.

At the beginning of this decade we began to hear about a possible Russian ABM system, and we became concerned about its potential effects on our ICBM and Polaris systems. In response the MIRV concept was invented. Today there are additional justifications for MIRV besides penetration, but that is how it started. Now, the possibility of a Russian MIRV is used as one of the main arguments in support of the Safeguard system. Thus we have come one full turn around the arms-race spiral. No one in 1960 and 1961 thought through the potential destabilizing effects of multiple warheads, and certainly no one predicted, or even could have predicted, that the inexorable logic of the arms race would carry us directly from Russian talk in 1960 about defending Moscow against missiles to a requirement for hard-point defense of offensive-missile sites in the U.S. in 1969.

By the same token I am sure the Russians did not foresee the large increase in deployed U.S. warheads that will ultimately result from their ABM deployment and that made it so counterproductive. Similarly, no one today can describe in detail the chain reaction the Safeguard deployment would lead to, but it is easy to see the seeds of a future acceleration of the arms race in the Nixon Administration's Safeguard proposal. Soon after Safeguard is started (let us assume for now that it will be) Russian offense planners are going to look at it and say something such as: "It

Figure 7. First Salvo launch of Minuteman ICBM's was made at Vandenberg Air Force Base in California on February 24, 1966. Photograph was released by U.S. Air Force.

may not work, but we must be prudent and assume it will." They may then plan further deployments, or more complex penetration systems, or maybe they will go to more dangerous systems such as bombs in orbit. A little later, when some of our optimistic statements about how "it will do the job it is supposed to do" have become part of history, our strategic planners are going to look at Safeguard and say something such as: "Maybe it will work as they said, but we must be prudent and assume it will not and besides, now look at what the Russians are doing.

The approach to strategic thinking, known in the trade as "worst-case analysis," leads to a completely hopeless situation in which there is no possibility of achieving a state of affairs that both sides would consider as constituting parity. Unless the arms race is stopped by political action outside the two defense establishments, I feel reasonably sure there will be another "crash program" response analogous to what we had in the days of the "missile gap"—a situation some would like to see repeated.

I also mentioned in my own testimony at the ABM hearings that "we may further expect deployment of these ABM systems to lead to the persistent query 'But how do you know it really works?' and thus to increase the pressures against the current limited nuclear-test ban as well as to work against amplifying it." I mentioned this then, and I mention it again now, in the hope that it will become a self-defeating prediction. It is also important to note that the response of our own defense establishment to the Russian ABM deployment, which I have outlined above, was not the result of our being "provoked," and I emphasize this because we hear so much discussion about what is a "provocative" move and what is not. Rather, our response was motivated by a deep-seated belief that the only appropriate response to any new technical development on the other side is further technical complexity of our own. The arms race is not so much a series of political provocations followed by hot emotional reactions as it is a series of technical challenges followed by cool, calculated responses in the form of evermore costly, more complex and more fully automatic devices. I believe this endless, seemingly uncontrollable process was one of the principal factors President Eisenhower had in mind when he made his other (usually forgotten) warning: "We must be alert to the...danger that public policy could itself become the captive of a scientific-technological elite." He placed this other warning, also from his farewell address, on the same level as the much more familiar comment about the military-industrial complex.

Several alternative approaches to Safeguard for protecting Minuteman have been discussed recently. These include superhardening, proliferation, a "shell game" in which there are more silos than missiles, and land-mobile missiles. Although I was personally hopeful before the hearings that at least one of these approaches would maintain its invulnerability, a review of the recent debates leaves me now with the pessimistic view that none of them holds much promise beyond the next 10 years.

Silo-hardening most probably does work now, in the sense that the combination of SS-11 accuracy and yield and Minuteman silo-hardening works out in such a way that one incoming warhead (and hence one SS-11 missile) has less than a 50-50 chance of destroying a Minuteman. If one considers the technological trends in hardening, yield per unit weight, MIRVing and accuracy, however, it does seem convincing that this is a game in which the offense eventually will win. Albert Wohlstetter, testifying in favor of the Safeguard system before the Senate Armed Services Committee, quoted a paper he wrote with Fred Hoffman in 1954 (long before any ICBM's were actually in place anywhere) predicting that the

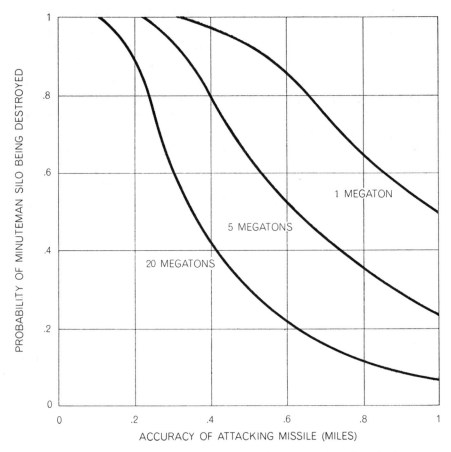

Figure 8. Vulnerability of minuteman is revealed in this graph, which relates probability of destruction of a hardened Minuteman silo to accuracy for three different sizes of attacking warhead. This graph was interpreted by Deputy Secretary Packard as demonstrating the seriousness of the threat to Minuteman posed by the large Russian SS-9 missile, which he said is capable of carrying either one 20-megaton warhead or three five-megaton warheads.

ability of silo-hardening to protect offensive missiles would run out by the end of the 1960's. That was a remarkably prescient study and is wrong only in numerical detail.

If we take the same rosy view of technology that was taken in almost all the pro-ABM arguments, then hardening will not work for more than another five years. My own view of the technological future is clearly much less rosy, but I do believe that the situation in which hardening is no longer the answer could come by, say, 1980 or, more appropriately, 1984.

Proliferation of Minuteman would have worked in the absence of MIRV. Now, however, it would seem that the ability to MIRV, which no doubt can eventually be carried much further than the fewfold MIRV we see for the immediate future, clearly makes proliferation a losing game as well as the dangerous one it always was.

The "shell game" has not in my view been analyzed in satisfactory detail, but it would appear to have a serious destabilizing effect on the arms race. Schemes have been suggested for verifying that a certain fraction of the missile holes are in fact empty, but one can foresee a growing and persistent belief on each side that the "other missiles" must be hidden somewhere.

Road-mobile and rail-mobile versions of Minuteman have been seriously studied for well over a decade. These ideas have always foundered on two basic difficulties: (1) Such systems are inherently soft and hence can be attacked by large warheads without precise knowledge of where they are, and (2) railroads and highways all pass through population centers, and large political and social problems seem unavoidable.

Where does all this leave us insofar as finding a technical solution for protecting Minuteman is concerned? One and only one technically viable solution seems to have emerged for the long run: Launch on warning. Such an idea has been considered seriously by some politicians, some technical men and some military officers. Launch on warning could either be managed entirely by automatic devices, or the command and control system could be such as to require authorization to launch by some very high human authority.

In the case of the first alternative, people who think about such things envision a system consisting of probably two types of detection device that could, in principle, determine that a massive launch had been made and then somewhat later determine that such a launch consisted of multiple warheads aimed at our missile-silo fields. This information would be processed by a computer, which would then launch the Minutemen so that the incoming missiles would find only empty holes; consequently the Minutemen would be able to carry out their mission of revenge. Thus the steady advance of arms technology may not be leading us to the ultimate weapon but rather to the ultimate absurdity: a completely automatic system for deciding whether or not doomsday has arrived.

To me such an approach to the problem is politically and morally unacceptable, and if it really is the only approach, then clearly we have been considering the wrong problem. Instead of asking how Minuteman can be protected, we should asking what the alternatives to Minuteman are. Evidently most other people also find such an idea unacceptable. As I mentioned above, the Army has found it necessary to reassure people repeat-

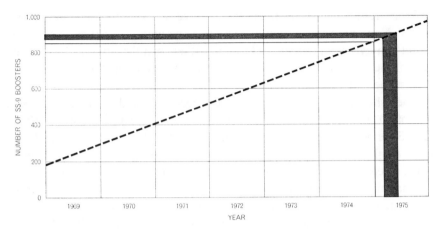

Figure 9. Safeguard could be nullified within a few months after its Phase I deployment, according to this graph, which is based on calculations presented by George W. Rathjens in his testimony before the Senate Armed Services Committee on April 23, 1969. His analysis took as a basis of calculation the implication in Secretary Laird's testimony that the Minuteman forces may become seriously imperiled in the mid-1970's. Assuming that the Russian SS-9's would have four or five MIRV warheads each by that time, Rathjens then estimated that approximately 850 SS-9's would have to be deployed in order to achieve this result. From this number, and the estimate of the current number of SS-9's deployed (about 200), he got a rate of deployment (about 100 per year). Making certain assumptions about the numbers and effectiveness of the Spartan and Sprint ABM missiles that would prolong the SS-9 production program by two to five months the Russians would be able to cope with Safeguard by simply exhausting it and would still have enough warheads to imperil Minuteman. Recently different numerical assumptions ahve been made, but they do not change the general conclusion that the proposed Safeguard system is much too thin to safeguard Minuteman.

edly that ABM missiles would not be launched without approval by "the highest authorities," even though this is clearly a far less serious matter in the case of the ABM missiles than in the case of Minuteman.

The alternative is to require that a human decision-maker, at the level of "the highest authorities," be introduced into the decision-making loop. But is this really satisfactory? We would be asking that a human being make, in just a few minutes, a decision to utterly destroy another country. (After all, there would be no point in firing at their empty silos.) If, for any reason whatever, he was responding to a false alarm, or to some kind of smaller, perhaps "accidental," attack, he would be ensuring that a massive deliberate attack on us would take place moments later. Considering the shortness of the time, the complexity of the information and the awesomeness of the moment, the President would himself have to be properly preprogrammed in order to make such a decision.

Those who argue that the Command and Control System is perfect or perfectable forget that human beings are not. If forced to choose, I would prefer a preprogrammed President to a computer when it came to deciding whether or not doomsday had arrived, but again I feel that this solution too is really unacceptable, and that once again, in attempting to defend Minuteman, we are simply dealing with the wrong problem. For the present it would seem the Polarises and the bombers are not, as systems, subject to the same objections, since there are now enough other approaches to the problem of ensuring their invulnerability to sudden massive destruction.

In my view, all the above once again confirms the utter futility of attempting to achieve national security through military technology alone. We must look elsewhere. Fortunately an opportunity does seem to be in the offing. There appears to be real promise that serious strategic-arms limitation talks will begin soon. The time is propitious. There is in the land a fairly widespread doubt about the strictly military approach to security problems, and even military-minded politicians are genuinely interested in exploring other possibilities. The essay by Academician Sakharov, as well as the statements of Russian officials, indicate genuine interest on the other side. The time is propitious in another sense: both sides will be discussing the matter from a position of parity. Moreover, this parity seems reasonably stable and likely to endure for several years.

Later, however, major deployments of sophisticated ABM systems and, even more important, widespread conversion of present single-warhead systems to MIRV will be strongly destabilizing and will at least give the impression that parity is about to be upset. If so, the motto of the arms race, "Let us err on the side of military safety," will come to dominate the scene on both sides and the present opportunity will be lost. Therefore in the short run we must do everything possible to ensure that the talks not only start but also succeed. Although the ABM decision may not forestall the talks, it would seem that success will be more likely if we avoid starting things that history has shown are difficult to stop once they are started.

Such things surely include deployment of ABM missiles and MIRV's. There have been successes in stopping programs while they were in the development phase, but seldom has anything been stopped after deployment had stared. The idea of a freeze on deployment of new weapons systems at this time and for these reasons is fairly widespread already, but achieving it will require concerted action by those believing strongly in the validity and necessity of arms limitations as a means of increasing national security. Thus the principal result of the recent national debate over the ABM issue has been to make it clear that Safeguard will safeguard nothing, and the right step for the immediate future is doing whatever is necessary (such as freezing present deployments and developments) to ensure the success of the coming strategic-arms-limitations talks.

In addition, the ABM debate has served to highlight more serious issues (for example the implications of MIRV for the arms race) and to raise serious questions about other weapons systems. For instance, I suggest that we have also found that silo-based missiles will become obsolete. The only sure method for defense of Minuteman beyond, say, the mid-1970's seems to be the unacceptable launch on warning. As long as we must have a strategic deterrent, we must find one that does not force us to turn the final decision over to either a computer or a preprogrammed President. Minuteman was conceived in the 1950's and served its purpose as a deterrent through the 1960's, but it appears that in the 1970's its threat to us will exceed its value, and that it and other silo-based missiles will have to go. The deterrent must have alternatives other than "go/no-go," and for the 1970's at least it would now appear that other strategic weapons (Polaris/Poseidon and bombers) could provide them. I expect, however, that as the continuing national debate subjects the whole matter of strategic arms to further public scrutiny we shall learn that these other alternatives also have dangerous flaws, and we shall see confirmed the idea that there is no technical solution to the dilemma of the steady decrease in our national security that has for more than 20 years accompanied the steady increase in our military power.

Arms-Limitation Strategies

In 1934, when the Nazi terror was just beginning, and artificial radioactivity and the neutron had just been discovered, Leo Szilard invented the idea of a neutron chain reaction. He believed there must be some atomic nucleus that would, upon absorbing one neutron, emit two neutrons and energy, and that these two neutrons would lead to four, and so on. He did not know what nuclear process would be involved or what element would support that process. Nevertheless, he quickly went on to sketch two distinct devices based on such a neutron chain reaction.

One of these devices would release energy in a steady and controlled fashion, resembling in principle today's nuclear reactors; the other would release energy suddenly and in an uncontrolled fashion, similar in principle to today's nuclear bombs. Szilard's intuition told him that the former would be a boon to mankind, and that the latter would constitute a most serious threat. His views concerning this threat were strongly influenced by the rise of Nazi terror and his projection of how the political situation would develop. (Szilard had only recently arrived in England as one of the very first of what was to become a flood of refugees.) His solution to the dilemma inherent in all this was classic: He took out a public patent on the reactor-like device and he managed, with some difficulty, to obtain a secret patent on the bomb-like device.

Thus, one might say secrecy was the first nuclear-arms limitation strategy. Indeed, governments still consider secrecy useful, although the approach to arms control they now favor most is the treaty. But because treaties to date have fallen far short of eliminating nuclear weapons or removing nuclear weapons from the realm of national sovereignty, they are, at best, only partial measures against the threat that Szilard foresaw. Nevertheless, these partial measures may buy time, so it is important that physicists who wish to participate in research and education concerning the prevention of nuclear war know about the measures that have been taken, and about those being proposed or negotiated. My goal here is to present some of this information, emphasizing international agreements that bear on arms control. Some readers will probably be surprised to find that there are quite a few relevant treaties in force or in various stages of development.[1]

Even since Szilard's secret patent, governments have used secrecy to attempt to separate the development and promotion of nuclear power from the development and acquisition of nuclear weapons. In particular, they have used secrecy to prevent the spread of nuclear weapons to other powers and, more recently, even to subnational groups. Among governments, secrecy is not only a highly regarded means to this end, it is the only universally agreed-upon means. Obviously, secrecy does not "work" in the absolute sense—weapons have already spread to other countries and it seems certain that they will continue to do so. However, secrecy does inhibit the whole process by sharply limiting the total number of people—scientists and inventors—who can engage in the kind of personal interaction that speeds the development of this or any technology, and it cuts off the interaction between groups of such people in different countries. Indeed, Edward Teller's well-known advocacy of reduction in the secrecy barrier is based in part on his belief that with less secrecy qualitative improvements in weaponry would come at a faster pace, and that would be a net benefit to the United States. In the early years, the American and Soviet programs did, in fact, unavoidably trade information by providing each other with radioactive fallout to analyze; but since the atmospheric test ban of 1963, that channel of communication has been closed.

Peril Anticipated

In December of 1938, Otto Hahn and Friz Strassman in Germany discovered that neutrons induced in uranium what was soon named the fission process. Other experiments and theoretical elucidations quickly followed, and it was immediately and widely recognized that here was the process that would make the neutron chain reaction possible in both its controlled and violent forms.

Programs to explore and, if possible, exploit these new discoveries started in at least six countries. However, only the American program was to succeed. Only in America was it possible, given the violent circumstances then prevailing most everywhere else, to marshall the intellectual and material resources necessary to produce an atomic bomb.

But, even before the Manhattan Project achieved its goal, many among its leadership were able to find some time to contemplate the future they were striving to bring about. They again took it as almost axiomatic that nuclear power would be a boon to mankind, and that nuclear research in general had an exciting and fruitful future before it. Many of them also believed that no matter what short-term benefits their work might bring, in the long run the advent of nuclear weapons, and the nuclear arms race that they feared would build up around them, constituted a most serious peril for the future of man and his civilization. President Harry Truman, and many other knowledgeable statesmen, soon came to share similar views.

Immediately after the war, as one of its responses to these growing concerns, the government established a special committee to explore means and generate proposals for coping with this great problem. David Lilienthal chaired the committee, which reported to the highest levels of government. Members included Robert Oppenheimer, Chester Barnard, Charles Thomas and Harry Winne. Because of his knowledge and intellect, Oppenheimer was the committee's central figure, just as he was in other similar instances at the time.

The Radical Solution

The Lilienthal Committee considered the general situation to be unprecedented both in its seriousness and its substance and they proposed what they held to be an appropriately novel and radical solution. They proposed to create an international agency that would "conduct all intrinsically dangerous operations in the nuclear field, with individual nations and their citizens free also to conduct, under license and a minimum of inspection, all non-dangerous, or safe, operations." They also proposed to eliminate nuclear weapons—eventually—saying that nothing less would suffice. In sum, they said the rules and customs that derive from the concept of national sovereignty must not be allowed to prevail in the realm of nuclear fission and all the processes and activities that derive from it.

It was not to be. The American plan, which was presented to the United Nations in 1946 by Bernard Baruch, and whose central substance consisted of the proposals of the Lilienthal Committee, was promptly rejected by the Soviets. Also, it seems unlikely that the American body politic would have accepted it in the end, even though it had the full blessing of the President himself.

Historians and analysts have often questioned the sincerity of Oppenheimer and Lilienthal, Truman and Dean Acheson, and Baruch. Surely, it has been said, they must have known the plan was too radical to succeed. In fact, the authors of the plan anticipated just such a reaction. In the report itself they wrote:

> The program we propose will undoubtedly arouse skepticism when it is first considered. It did among us, but thought and discussion have converted us.
>
> It may seem too idealistic. It seems time we endeavor to bring some of our expressed ideals into being.
>
> It may seem too radical, too advanced, too much beyond human experience. All these terms apply with peculiar fitness to the atomic bomb.
>
> In considering the plan, as inevitable doubts arise as to its acceptability, one should ask oneself, "What are the alternatives?" We have, and we find no tolerable answer.

I believe Oppenheimer and the others were right: There are no alternative solutions to the problems created by the advent of nuclear weapons. The other approaches discussed in the remainder of this article are all palliatives, means for moderating or delaying the nuclear holocaust, but inadequate for preventing it altogether. Of course, in a situation as dreadful as the one in which we find ourselves, it is very definitely worthwhile to pursue and promote palliative and partial measures in the hope that they will both buy time and eventually lead to a real solution. But while we are pursuing such measures, we should bear in mind their true nature.

During the first decade following World War II, no progress toward arms control was possible, but in the mid-fifties the outlook improved. A number of factors contributed importantly, including the achievement of the atomic bomb by the Soviets, a development that made it possible for them to deal with the West at least nominally from a position of equality. In addition, Stalin died and was replaced by leaders more open to dealing with the outside world. At about the same time, the international dialog, which had never totally ceased, changed from one with a strong rhetorical emphasis on total measures—"general and complete disarmament"—to a more serious search for partial measures that could be taken up individually and in a serious manner more likely to lead to success in each case.

Since then, governments have proposed and attempted many such partial measures, and some have been achieved. These measures have come in several quite different formats and have involved a wide variety of substantive issues. The formats have included bilateral treaties such as SALT, multilateral treaties such as the Nonproliferation Treaty, United Nations

Conventions such as the Convention on Biological Warfare, carefully matched unilateral actions such as the test moratorium of 1958-61, simple unilateral actions such as the 20 years of US restraint in the development of antisatellite weapons, and the establishment of special international regimes such as the International Atomic Energy Agency. The substance of these measures has varied even more widely, and has included a variety of limitation on nuclear testing, restrictions on providing assistance to others, prohibitions on the misuse of assistance from others, the denuclearization of specific regions such as Latin America and outer space, numerical ceilings on specific deployments, the legitimization of certain unilateral intelligence-gathering techniques and the establishment of some cooperative means of verification.

Treaties Now in Force

The arms-limitation format most favored by governments, statesmen and diplomats is the treaty. The principal advantage of a treaty over, say, paired unilateral actions, is that it spells out in detail all the limitations and restrictions that are to be undertaken, the means to be used for verifying compliance, the duration, and the conditions under which it may be terminated. Only by providing such details is it possible to avoid the misunderstandings (which often seem deliberate) that so commonly characterize relations between states with strongly differing political systems. Moreover, treaties between the US and the USSR are developed step by step and in detail by the superpowers themselves. As a result, all of their provisions are carefully tailored to be in the mutual interest of both parties, thus ensuring a higher probability of compliance than in the case of UN conventions and other very broadly international arrangements.

Conversely, the principal disadvantage of the treaty approach is that negotiation and ratification often take a very long time, during which external events can take place that cause the process to abort. One case in point is SALT II, which was slowed and delayed by a series of such events— Deng Xiao-Ping's visit to Washington, the controversy over the brigade of Soviet troops in Cuba, the Teheran embassy capture—and finally aborted by the invasion of Afghanistan. Another case is that of the Comprehensive Test Ban Treaty negotiations during the Carter Administration. Internal opponents of the treaty were able to search out and use bureaucratic maneuvers to slow the negotiating process until it too was finally aborted by the same events that killed SALT II.

Let us take a quick look at some of the arms-control treaties now in force.

- The Antarctic Treaty of 1959 in essence demilitarizes Antarctica. It was signed by all the parties having territorial claims or other special interests in the region. A special feature is its provision for on-site inspections of the various research bases maintained in the region to assure that they are not conducting banned activities. The United States has frequently exercised its rights under this provision.
- The Limited Test Ban Treaty of 1963 prohibits nuclear testing in the atmosphere, outer space and under water. The original intent was to ban nuclear tests altogether, but in the end the US view was that no adequate means were available for verifying a ban on underground tests so these could not be included in the treaty. Numerically speaking, most of the public support for this treaty came from those whose principal stated concern was the health problem posed by radioactive fallout, and their concern was, of course, satisfied by the limited test ban.
- The Outer-Space Treaty of 1967 and the Seabeds Treaty of 1972 ban the placement of "weapons of mass destruction" in those two locations.
- The Nonproliferation Treaty of 1968, which went into force in 1970, is designed principally to bar the nuclear-weapons states from helping other states to get or to build nuclear weapons, and to bar the non-nuclear-weapons states from seeking to obtain or build such weapons. The treaty contains several additional provisions designed specifically to make it more palatable to the nuclear have-nots. One such provision calls for the nuclear-weapons states to "pursue negotiations in good faith on effective measures relative to cessation of the nuclear arms race." Others call for the nuclear-weapons states to help the non-nuclear-weapons states acquire the benefits of nuclear energy, including benefits deriving from the so-called "peaceful uses" of nuclear explosives. The treaty calls for a review of the nonproliferation regime at five-year intervals. In the past two such reviews, there have been widespread charges that the superpowers have not lived up to their obligations under all of the additional provisions.
- The Latin American Nuclear Free Zone Treaty of 1968 is designed to eliminate all nuclear weapons from Latin America. It has been ratified by most but not all of the states of the region. It also includes protocols placing certain consistent requirements on states outside

the region when those states own Latin American territories or posses nuclear weapons. This treaty has served as a model for as yet unfulfilled proposals for "nuclear-free zones" in several other parts of the world.

- The Antiballistic Missile Treaty was ratified in 1972. Together with its later amendments, this treaty defines an ABM system and places numerical and geographical limits on its deployment. It contains provisions designed to limit certain characteristics of ABM subsystems and to inhibit, but not prevent, the development of new types of ABMs. In addition, it contains provisions legitimizing and protecting so-called "national technical means" of verification and restricts practices that would confound such systems. The treaty is of indefinite duration, but it is to be reviewed every five years. It also calls for the establishment of a "Standing Consultative Commission," which would meet in private and consider complaints and questions raised by either side. This commission has met as required and has reported that all questions raised have been resolved satisfactorily.

- The Executive Agreement Covering Certain Offensive Systems, also ratified in 1972, places limits on the numbers and sizes of specific offensive systems, including ICBMs and sea-launched ballistic missiles. This agreement was in essence a "freeze," because the limits were set at the numbers of systems already deployed or in the process of being deployed. It restated the provisions dealing with national technical means of verification. It was to run for five years, at which time it was to be replaced by a treaty that would be broader and more far reaching in scope. The United States and the Soviet Union have not achieved this, but through a continuing series of understandings they have kept the executive agreement in force.

Uncompleted Treaties

The SALT II Treaty was negotiated and signed during the Carter Administration. It was called for in the executive agreement of SALT I, and was based in large part on certain general "guidelines" presented in the Vladivostok agreement worked out by Presidents Ford and Brezhnev in 1974. It sets numerical limits on all the major types of strategic offensive systems, and it sets sublimits that restrain the total number of warheads by limiting the "MIRVing" of delivery vehicles. The treaty contains a data base and numerous working definitions, all of which are designed to make it easier to achieve further agreements, which the treaty tacitly assumes will follow, and which would eventually lead to a real reduction in numbers. The SALT II treaty was finally signed and submitted to the Senate in late 1979 after several long delays. But before it could be ratified, the events mentioned above intervened and put the necessary two-thirds majority in the Senate beyond reach.

The Threshold Treaty of 1974 and the Peaceful Nuclear Explosions Treaty of 1976 were worked out and signed in the Nixon and Ford Administrations respectively. The first was designed to prohibit tests of nuclear weapons exceeding 150 kilotons in yield, and the latter was designed to allow multiple nuclear explosions conducted for peaceful purposes to exceed this limit in the aggregate but not individually. Two very important features of these treaties are that they provide for an exchange of geophysical data concerning the test site, and that they call for on-site observations and measurements by personnel of the other principal in certain special circumstances. This was the first instance in which the Soviets ratified a treaty allowing such intrusive procedures to take place on their home territory, but these treaties have not been ratified by the United States, so these special provisions are not in force. Both states, however, have said they would for now comply with the 150-kt limit, and both appear to be doing so.

Treaties attempted or proposed

A Comprehensive Test Ban, which would prohibit all nuclear weapons tests everywhere, has been a stated goal of both the US and the USSR since 1958. The pursuit of this goal has led to several partial measures—the Limited Test Ban and Nonproliferation treaties, and the signed-but-not-ratified Threshold and Peaceful Nuclear Explosions treaties—but the goal itself remains elusive. Negotiations during the Carter Administration did achieve several advances of fundamental importance in the area of verification. The US and USSR reached agreement in principle on a system of voluntary on-site inspections. These would be based on a series of challenges and required responses and the deployment of national seismic stations (the "black boxes" of earlier years), which would be placed on each other's territory and which would be constructed and operated so as to provide a continuous, unadulterated, stream of data in near real time. In addition, the Soviets agreed to forego at least temporarily their program to develop and use so-called "economic nuclear explosions," analogous to our "peaceful nuclear explosions." After some important early progress, this negotiation encountered a steadily and seriously deteriorating political climate (Teheran, Kabul, the 1980 US elections) and was eventually stalled by the bureaucratic maneuvers of its opponents. As a result, the details underlying these agreements-in-principle were never fully elaborated. When President Reagan took office in 1981, the negotiations were cancelled.

The anti-satellite, or ASAT, negotiations of 1978-79 were intended to forestall the construction of anti-satellite weapons systems and the eventual preparation to conduct warfare of various novel kinds in space. Three negotiating sessions took place during the early Carter years, but it eventually became evident they were leading nowhere and they were abandoned even before the international scene seriously deteriorated. In my view, this happened largely because neither side was able to develop a clear and generally internally acceptable view of its objectives.

During the last quarter century, several proposals to limit certain nuclear weapons activities have been put forward but never seriously negotiated. These included proposals to cut off production of the special materials needed to construct nuclear weapons, to limit seriously launches of large rockets so as to inhibit greatly their further development, and to nip in the bud the MIRV development program. In addition, there have been proposals for many other "nuclear-free zones," as well as proposals for making certain parts of the ocean safe "havens" for nuclear submarines as a means of supporting the so-called mutually assured destruction, or "MAD" doctrine.

Conventions and the like
United Nations conventions have the advantage of being universal, or nearly so. And, to the extent that they work and exercise a positive influence, they tend to reinforce other peacekeeping and peace-supporting activities of the UN. The corresponding disadvantage, as compared to bilateral treaties between states of roughly equal size and power (or multilateral treaties between two distinct sides or two dominant parties), is that they are not taken quite so seriously. Thus, while the record of compliance with arms-control treaties is generally quite good, the record in the case of these more broadly based arrangements is quite different, as evidenced by the many apparent violations of the Chemical Protocol and the Biological Convention, and the cavalier way in which many states, the US and USSR included, fail to live up fully to the purposes and regulations of the International Atomic Energy Agency. The reason for this difference, I believe, lies in the fact that the superpowers have must less control over the details of universal conventions, and they frequently end up facing a "take it or leave it" situation. They often choose to "take it," but

then are less likely comply with the detailed provisions than when they are in full control of the details. I believe Soviet behavior in connection with the Chemical and Biological Conventions is a case in point. Both are broadly multilateral conventions produced by an international organization, not by direct principal-to-principal negotiations, and the latter in fact contains no provisions for verification.

The 1925 Geneva Protocol on Chemical Weapons and the 1977 Biological Weapons Convention were sponsored by the League of Nations and the United Nations, respectively. The former seeks to prohibit the use of poisonous gases and "bacteriological methods" of warfare, and the latter seeks to prohibit the development, production and stockpiling of bacteriological and toxic weapons as well. The Geneva Protocol went into effect in 1928, but the United States did not ratify it until 1975, when it also ratified the Biological Weapons Convention. The parties to the latter also agreed to destroy such stocks as then existed within nine months after it went into force.

The International Atomic Energy Agency, which was created under the aegis of the UN, grew out of President Eisenhower's "Atoms for Peace" speech of 1953. Its twin purposes were to promote the spread of nuclear energy throughout the world and at the same time prevent the spread of nuclear weapons technology along with it. While the system of inspections and other procedures cannot prevent the spread of nuclear weapons—as those knowledgeable of but hostile to the Agency like to repeat— this system can and does greatly inhibit the process. Similar but more limited arrangements with the same objective have also been carried out in other contexts. For instance, the European Atomic Energy Community has similar purposes within Europe.

Unilateral Actions

The defense programs of all countries involve a continuous stream of unilateral actions, many of which have the effect of moderating the nuclear arms race. But nearly all of the moderating actions are the result of fiscal, not political, restraints. Occasionally, however, a unilateral restraint is imposed for the sole or principal purpose of moderating the military confrontation or avoiding some particular exacerbation of it.

A particularly interesting and important example was the 20-year period of restraint imposed by US presidents on the development of anti-satellite weapons. The United States, almost from the very beginning of the space age, has used its military space assets—reconnaissance satellites, for example—for a variety of very important national-security purposes. From Eisenhower onward, our Presidents have concluded that we would derive a net benefit from a situation in which no state had any anti-satellite capability. The US, therefore, was willing to forego a program to develop such weapons, in the hope that the Soviets would follow suit. As a result, proposals rising in the US Air Force and aerospace industry to develop a general-purpose anti-satellite weapon were continually rejected.

In the meantime, and despite US restraint, the Soviets launched their first anti-satellite experiment in 1968, and they have continued since then to conduct further tests in an on-again off-again fashion. Finally, in 1977 President Carter decided to make a three-pronged response to those Soviet actions: first, to initiate the development of a US anti-satellite weapon; second, to explore means for defending satellites against attack; and third, to begin negotiations with the Soviets to forestall all such developments. Beginning at Helsinki in 1978, three negotiating sessions took place, and, as I mentioned above, they got nowhere. Now both the US and the Soviets have anti-satellite programs underway.

Another important unilateral action took place in October 1958 when, after a succession of slowly converging unilateral statements by Eisenhower and Khrushchev, the US and the Soviet Union both suspended nuclear tests. This moratorium, as it was called, was in essence based on a matched set of unilateral statements in which each party pledged not to test if the other party did not test. The purpose was to create an atmosphere suitable for working out a formal treaty on the subject, and Eisenhower estimated that one year should suffice. The negotiations did not work out, however, and after fourteen months, on 29 December 1959, Eisenhower denounced the moratorium, saying the US was no longer bound by it, but would not begin testing without giving notice. Three days later, the Soviets denounced it also, but said they would not test unless the "Western powers" did so first.

When France tested its first nuclear weapon three months later, Khrushchev took formal and public note of it. Hence, as of that time there was no longer any *de jure* basis for a nuclear moratorium, but testing did not in fact resume until 18 months later, when an extensive Soviet series put a final end to the then purely *de facto* moratorium. Although there were no external legal restraints—formal or informal—on such testing by any party at that time, US official opinion has always held that we were in an important sense entrapped by the Russians. In the words of one current high-level White House official, "they used the moratorium to gain a full lap on us" in the cycle of nuclear testing and development. Ever since then, unverifiable moratoria, whether bilateral or not, have found little support in US government circles.

In recent years, the Chinese and the Soviets have made unilateral pledges to the effect that they would not be the first to use nuclear weapons, and many have urged that the United States also make such a "no first use" pledge. United States policy has always been to reserve the right to initiate the use of nuclear weapons, particularly in situations where the conventional balance may be very heavily weighted against us, and so we have always declined to join in such a pledge, whether on a unilateral or a multilateral basis. Another basic argument has been that such a pledge, if unaccompanied by any further, more concrete actions that might really make first use more difficult or less desirable, would be too easily reversed or disregarded.

More recently, a number of Europeans and Americans have proposed that we unilaterally and as a "first step" either eliminate all so-called "battlefield nuclear weapons" (such as nuclear artillery) or at least remove them from close proximity to the inter-German border. They argue that the actual use of such weapons for resisting an armed attack would not be to our net advantage, that we can successfully develop and deploy conventional means for resisting such an attack, and that removing such battlefield "nukes" would substantially reduce the probability of nuclear war in Europe. They argue further that the longer-range nuclear systems, which would presumably still remain available after such a "first-step," would be fully adequate for continuing the state of nuclear deterrence that has persisted the last 35 years.

The Freeze Movement

The people behind the current wave of "nuclear freeze" proposals have two distinct purposes. One is to provide a simple, easily understandable focus for public opinion; the other is to outline the substance for an international agreement to stop the nuclear arms race in its tracks.

In their first role, the proposals have taken the form of resolutions placed before the Congress and before the voters in many state and even local elections. In their second role, the goal is usually said to be a "bilateral and verifiable" freeze in the development, production and deployment of all nuclear weapons systems. Only through the treaty-writing process discussed above would it be possible to elaborate and negotiate the details necessary to achieve and sustain such a goal. A number of treaties encompassing partial freezes have, in fact, already been fully or partially negotiated—SALT I, SALT II, the Comprehensive Test Ban—so there are some precedents for such an action. However, some of the experts who commonly advocate nuclear restraints in general have argued that the freeze proposals include too many elements that are too widely disparate to serve as the basis for a unitary bilateral or multilateral negotiation.

It is clear that in its first purpose, the freeze movement is succeeding. In November 1982, in eight of the nine states where a freeze resolution appeared on the ballot, the electorate favored it. I believe these and similar actions will stimulate some real progress by governments in moderating the nuclear confrontation and achieving formal agreements further limiting the development and deployment of nuclear weapons, especially in Europe. However, it is not at all clear to me at this time precisely what final and formal form these new limitations will take. Nor is it clear whether any such progress can happen soon, or whether it must await the arrival of an administration in Washington (and perhaps in Moscow) more closely attuned to the public's concerns in this vital matter.

Reference

1. Arms Control and Disarmament Agreements, 1982 edition, United States Arms Control and Disarmament Agency, Washington, D.C. (1982).

Thinking About the Arms Race

My message can be summed up in a simple phrase: Zeal is not enough. The nuclear arms race constitutes one of the most dangerous and difficult problems that mankind has ever had to face. As with most such problems it is necessary to understand it in some detail in order to deal with it. If you want to be more than just a spear carrier in an opera written by someone else, if you hope to make a substantial contribution yourself, then you must make the effort and take the time necessary to understand in a fairly thorough fashion at least one of the several key aspects of the problem.

A recognition of the seriousness of the problem, is, of course, essential, but in itself it is not enough. If it were we might be much further along in the solution of the problem, for the fact is that many world leaders have described it in properly awesome terms. Let me begin by quoting some of them.

On that morning in July 1945 when the desert at Alamogordo was lit for the first time ever by the light that was "brighter than a thousand suns," Robert Oppenheimer, the director of the Los Alamos laboratory where the first atomic bomb was assembled, was inspired to recall a bit of Hindu scripture in which Vishnu said "I am become Death, the Destroyer of Worlds." A few years later he remarked "In some sort of crude sense which no vulgarity, no humor, no overstatement can quite extinguish, the physicists have known sin."

At about the same time, Albert Einstein wrote: "The unleashed power of the atom has changed everything except our way of thinking. Thus we are drifting toward a catastrophe beyond conception."

Henry Stimson, the U.S. Secretary of War, who authorized the bombing of Hiroshima, shortly after that event said in a memo to President Truman: "I think the bomb constitutes merely a first step in a new control by man over the forces of nature too revolutionary and too dangerous to fit into the old concepts. I think it caps the race between man's growing technical power and his moral power."

General Douglas MacArthur, a soldier of great talent, and the author of a number of pithy quotations, one of the most famous of which is "In war there is no substitute for victory," later concluded that the atomic bomb had "destroyed the possibility of war being a medium of practical settlement of international differences."

President Eisenhower, reminiscing about his term in office, remarked that his greatest disappointment had been his failure to achieve a treaty banning all nuclear weapons tests forever.

Robert McNamara, Secretary of Defense under Presidents Kennedy and Johnson, spoke of the nuclear arms race as having "a mad momentum of its own."

Admiral Rickover, "father" of the nuclear submarine, said when he retired that he was not proud of the work he had done, but that it "was a necessary evil." He added that he thought mankind would destroy itself.

Pope John Paul II, referring to nuclear weapons in a speech before UNESCO, said "I speak to you in the name of this terrible menace which weighs on humanity...and I beseech you: Let us display all our efforts to instill and respect, in all the domains of science, the primacy of the ethical. Let us especially display our efforts to preserve the human family from nuclear war!"

President Reagan, early in 1983, said that "it is unthinkable...to look down on an endless future with both of us sitting here with these horrible missiles aimed at each other, and the only thing preventing a holocaust is just (that) no one pulls the trigger."

I think that last statement is just exactly "right on," as they say, even though I cannot take his particular solution to the problem at all seriously.

These statements might, at first glance, lull us into imagining that all that is needed to overcome the threat of nuclear war is to spread the message so that a majority of the population understands just how bad and dangerous the situation is. Then, somehow, it ought to take care of itself. Indeed, considering that many of the people whose views I have quoted have been leading figures in developing and carrying out national security policy, we might well wonder why they didn't fix the situation themselves when they had the opportunity to do so.

Now, I believe that all these statements are true and relevant, and that they correctly convey the danger and urgency of the matter. The reason that we cannot build a solution directed on them is, therefore, not that they are false or irrelevant, but that they are incomplete.

So, Why Is It So Hard?

The reason it has proven to be so hard to cope with this problem is, I believe, that we and our national security leaders face not one but two serious problems and that these problems are so very tightly intertwined that they must be faced and solved simultaneously. The popular shorthand name for one of these problems is "The Russian Threat" and the popular shorthand name for the other is "The Nuclear Arms Race." The first of these problems is of a more immediate nature and calls forth concrete proposals for addressing it. The second is of an ultimate nature and calls forth more abstract and general proposals for solving it. "The Russian Threat" emanates from a huge empire with a 400 year record of steady encroachment on its neighbors on all sides and presently equipped with, among other things, 40,000 tanks in East Germany and 10,000 nuclear warheads poised and ready, and less than one half hour from their targets in North America. The responses that it elicits involve forging alliances to contain any further Russian expansion, by its own or proxy forces, and the development, construction and deployment of conventional and nuclear weapons to counterbalance those the Russians possess. "The Nuclear Arms Race," on the other hand, threatens us and all of mankind with a nuclear holocaust on that indeterminant day when for whatever reason our policy of nuclear deterrence finally fails. The responses that it elicits include proposals to replace confrontation with detente, to negotiate treaties to control, limit, and eventually eliminate nuclear weapons and, more generally, to reinforce those international institutions that try to create and promote international law and provide at least a modicum of international law enforcement.

It is not just some experts and students of national and international security who see things this way. Studies of American public opinion clearly show that the views of the American public at large are much the same and involve very similar contradictions. Research shows that the opinions that the American public holds on specific security issues are built up out of five essential beliefs that have remained quite solid and stable over the years regardless of whatever immediate and short term problem may be engaging our attention at any particular moment. These beliefs are:

1. Nuclear war is a crazy idea.
2. A limited nuclear war will not stay limited.
3. The nuclear arms race is stupid and wrong, especially now when so many other needs are unmet. but!
4. We cannot trust the Russians.
5. We must maintain a military capability sufficient to defend ourselves and to meet our obligations to our allies.

The policy problem, then, is that we face not one, but two, critical issues simultaneously, and we and our leaders must search out and implement solutions to each of them that do not make the other one worse. That in turn means that many of the simple solutions that are commonly proposed won't do, but it does not mean that there are no solutions. We must, however, note that in such a situation, that is, in a situation in which the simplest, cheapest and most straightforward means for handling an immediate problem run counter to the means necessary for solving an ultimate problem, we human beings, whether acting either individually or collectively, have never done very well. Indeed, if we really knew how to behave when short term needs led us in one direction and long term needs in the other, no one would smoke, no one would be overweight, and there would be little crime.

The Soviet–American Confrontation

Let me examine first some of the specifics of the Soviet–American confrontation. In doing so, I hope to show that looking at it with some care will reveal a very different picture than that resulting from superficial analyses conducted from the perspective of either of the ideological extremes.

On the one hand we find President Reagan, in a speech before a group of religious fundamentalists, giving vent to deep feelings held by many other Americans, and Europeans as well, when he said "We are faced with an evil empire" which is the source of all the world's problems. Others who have a similar orientation like to quote Lenin when he talked about the inevitable world wide victory of Marxism, and they remind us that Lenin said we capitalists were so blinded by avarice that we would ultimately sell the communists the rope they would use to hang us.

On the other hand, the western left makes parallel statements and tells corresponding stories. Apologists for the Russians commonly insist that they are only behaving the way victims of decades of capitalist encirclement should be expected to behave. They remind us that the United States and other western nations sent troops into Russia in the aftermath of World War I in a futile attempt to stave off a communist victory. These notions are often extended to justify Russian political subjugation in this century of almost a dozen other formerly fully independent nations.

I believe that the current situation is much better understood if we avoid either "devil" theories or "mea culpa" apologetics.

The basic facts are that the current world system consists of more than

.150 independent states, that these states accept very little law defining what their relationships ought to be, and that there is absolutely no law enforcement governing their behavior towards each other. In brief, the relationships between nations today are essentially anarchistic and chaotic, as the Secretary General of the United Nations, Mr. Perez de Cuellar, has noted. This situation is the result of hundreds, even thousands of years of political evolution, influenced to a substantial degree by the technological changes that have taken place over that span of time. We have moved slowly and steadily from a simple situation in which small units related lawlessly, if at all, to other small units, to more complex situations in which progressively larger units behave towards each other in the same way, from the family to the tribe to the city-state to the nation-state, from strictly local scuffles to world wars. Perhaps the biggest difference between the earlier situation and the present one is that isolation has been replaced by interdependence, with the result that conflicts are more likely. The current circumstances give rise almost inevitably to conflicts that can only be resolved by force or the threat of it, and I venture to say that if neither the revolution of 1776 nor the revolution of 1917 had occurred, the world political scene would not be very different in this respect. De Tocqueville, more than a century before either the rise of Marxism–Leninism or the invention of the atomic bomb, predicted that Russia and America would be the superpowers of the twentieth century and that they would be rivals. He even implied that they would be committed to opposite ideologies.

Let me turn briefly to the military aspect of the current situation. When World War II broke out, just over forty years ago, both of the superpowers proved to be invulnerable. When Hitler attacked Russia in 1941, he succeeded in a matter of days in almost totally destroying the Red Army that faced him. All that then lay between the Germans and victory was distance, but that turned out to be enough. Before the German armies could conquer that, the Soviets managed to call up reserves, bring in distant troops, mobilize civilians, and get help from abroad. The sheer size of the Soviet Union had protected it from modern history's greatest war machine. And it was the same with us; distance, in the form of our two great ocean barriers, made us virtually invulnerable to external attack. With the acquisition of atomic bombs and intercontinental delivery systems by the United States, the Soviet situation changed drastically. By the late 50's the United States could have totally destroyed the Soviet Union and nothing the Soviets could have done, other than threaten revenge, could possibly have prevented it. But of course the United States' monopoly on nuclear weapons could not last, and by the mid 60's the Soviets had achieved a similar capability. Since then the situation has steadily become worse; each of us has built up still more powerful—that is, more destructive—military forces and each has, as a consequence, become less secure and more vulnerable, as measured by what the other side could do if it simply decided to attack and accept the consequences.

The engine of this great change, the specific origin of this unhappy state of mutual and total vulnerability, has been what most of us call the nuclear arms race. Each side has gone from zero nuclear warheads in 1945 to roughly 30,000 apiece in 1983, and each side has gone from no capability for the intercontinental delivery of any kind of weapon in 1945 to the deployment of forces that can move hydrogen bombs from their storage sites on one continent to their targets on another in less than 30 minutes. The course of these changes has sometimes been erratic. Sometimes one side would be building up rapidly while the other expanded more slowly or not at all, and sometimes the important changes have been in quality rather than in quantity. These variations in pace and style have led some analysts to claim that there really is no arms race or, at least, that only one side is racing, but the basic fact remains that we have both gone from nothing to something absolutely horrendous in just 40 years and we have both acted most of the time with a sense of great urgency. That situation and that behavior amount to what we call the nuclear arms race. Moreover, the number of nations participating in this race has grown already from one to five (with several additional ambiguous cases). It will surely grow beyond that in the future, and with that growth will inevitably come an increase in the number of ways that nuclear war can start either deliberately or accidentally.

Each of the superpowers has a deep distrust for the other. This fact makes the situation worse and at the same time makes it harder to do anything about it. And, although this distrust is mutual, the elements that make it up are quite different in the two cases.

On the one hand, as we all know, our distrust of the Soviets is based largely on the widely held belief that the Soviet leaders are both perfidious and expansionist, that is, that they will seize any opportunity to gain an advantage over us, to threaten our vital interests, to undermine our relations with others, and that they will do any or all of those things without any regard to prior promises or understandings. Any good "professional anticommunist" can give a long list of particulars, some of which are even real, and I will not now take the time for examples.

On the other hand, their distrust of us involves very different factors. In brief, Soviet officials regard us a unreliable and volatile. They complain that our behavior, and even our attitudes, are highly unpredictable and that we are liable to do almost anything in a sudden fit of pique. This view of us by Soviet leaders is based both on their inability to understand our system and on our actual behavior. As an example of the latter, consider our behavior as a negotiating partner. In the years between 1973 and 1983, three of the treaties we have spent years negotiating with the Soviets have been left in limbo at the last minute. In each of these cases, after signing the treaties, we then declined to finish the process by ratifying them. This matter is not all a partisan issue: one of these unratified treaties was signed by President Nixon (the TTBT), one by President Ford (the PNET) and one by President Carter (SALT II). Nor is it just a current problem. The same thing happened in the case of the agreement establishing the League of Nations in 1919 and with the Geneva Protocol on chemical warfare in 1923. (In the latter case, we actually did ratify it, but only 50 years later!) No other nations behave in this way, neither democratic nor autocratic ones, and we cause our friends as many problems in this regard as we cause our enemies. This erratic behavior arises partly as a result of our unique presidential system, but sudden changes in the US position often occur in the middle of a presidency as internal political winds buffet our leadership and cause sudden changes in the direction of international policy.

Another important asymmetry lies in the amount of information available about what is going on in each of the superpowers and in what the leadership in each case is thinking and saying. The United States is the most open country in the world, and the Soviet Union, at least among industrial powers, is the most closed. In the United States, some people actually make a good living by deliberately ferreting out and publicizing government secrets, and others—mostly in the government itself—further their own personal causes and careers by deliberately leaking secret information. In the USSR things could scarcely be more different. About one third of the territory of the USSR is officially off limits to foreigners, and nearly all the rest—perhaps as much as 99% of the total—is, as a practical matter, also closed to travel. On top of this there is government control of all media, political repression of all dissent, absolute intolerance for "whistle blowers," and a long tradition that everything is secret unless explicitly declared to be otherwise. Under such circumstances it is not just paranoia to wonder and worry about what might be going on over there that we don't know about that might be inimical to our interests. Indeed, in my view, this combination of three factors uniquely characterizing the Soviet Union—a tradition of secrecy, a repressive political system, and a huge and remote territory—represents, more than anything else, the chief obstacle we face as we try to find our way out of the current nuclear deadlock. Those American officials whose views about the Soviets are in-

formed largely by hatred and fear react to this real problem by grossly exaggerating the requirements for adequately monitoring formal agreements
with the Soviets. Thus, the whole process of negotiating agreements becomes much slower and more tendentious than is really necessary. Again,
this does not mean there is no way, but it does mean the problem is a lot
harder than it seems at first. There is a saying in international political circles that sums up the situation neatly. "The USSR," it is said, "is a country
of many secrets and few mysteries, whereas the USA is a country of many
mysteries but no secrets."

The great difference in the availability of information in the two countries leads to numerous errors in judgment at both ends of the western political spectrum about what is really going on in the USSR. On the one
hand, our leaders are continuously being badgered by the media and often
end up being trapped or goaded into making careless or casual remarks
which can easily be characterized as uninformed, chauvinistic, jingoistic
or just plain confused, and such statements are collected by our own internal political opposition and used to show how simplistic our current leaders are and how dangerous (or trivial) their "real" thinking is. By contrast,
Soviet leaders are never subjected to such pressures and they almost never
make casual remarks about anything, not even about their families. The
lack of such statements on their side often leads some of us, especially
those who tend to apologize for Soviet behavior, naively or erroneously to
conclude that the Soviet leaders are not as simplistic, dangerous, and uninformed in their approach to world affairs as are our own leaders. On the
other hand, on the relatively rare occasion when some Soviet colonel or
general publishes an article which may in fact be littler more than a pep
talk to the troops or a bit of whistling in the dark, our more jingoistic analysts will seize on it the way some anthropologists have been known to
make tenuous extrapolations from a humanoid jawbone. They will weigh
it for every possible nuance and derive from it proof that the Soviets have
a fully elaborated war fighting strategy and a practical plan for defending
their population and achieving victory in a nuclear war.

Hardware Issue

I hope I have said enough to indicate why we need a better appreciation of
the Soviet–American confrontation. Now let me suggest that we also need
a clearer understanding of what is involved in the nuclear arms race.

The burden imposed by distrust and misunderstanding is compounded by erroneous interpretations of the technical characteristics of the military hardware that figures in the nuclear arms race.

Very often the arguments made at both extremes of the political spectrum about particular nuclear weapons and the policies governing them are full of exaggerations, errors, and just plain nonsense. The debate about the so-called "neutron bomb" is a case in point.

At one extreme of this debate we find, among other publications, the magazine which boasts that it has the largest circulation and readership in the world: The Readers' Digest. Unhappily, in matters having to do with armaments in general and nuclear weapons in particular, The Readers' Digest usually takes a completely polemical point of view, that is, it consistently publishes articles that are not designed to inform but to persuade us to the publisher's predetermined point of view.

One particular article, published shortly before the 1982 elections and designed mainly to warn against the freeze movement by claiming that the KGB was behind it, included a substantial discussion of the neutron bomb. The article claimed that the KGB was particularly interested in stopping the deployment of this new weapon because it promised to play a vital role in stopping a Soviet invasion of western Europe. The article was artfully written so as to imply first that the neutron bomb was a completely new device with revolutionary characteristics, second, that these characteristics were exactly what was needed to stop a Soviet armored attack, and third, that we had no other practical alternative for accomplishing that result. Now, none of these implications are true, but most of the time the author of the article keeps his arguments fairly general, so it is hard to rebut them by challenging details, but in one place he does give some specific data which are supposed to compare the new bomb's characteristic effects with those of the bombs it is intended to replace, and when he does so he is wildly in error. He says in particular, "The smallest of the nuclear weapons then (1976) stored in Europe had a destructive force roughly equivalent to that of the bomb dropped on Hiroshima. The blast and heat from such a weapon would wipe out not only Soviet invaders but everybody and everything within a four mile radius of the detonation point." The introduction of the neutron bomb would change all that, the author claims, because "the ERW (the neutron bomb) obliterates everything within about 120 yards, inflicting no damage beyond that." This comparison is in error not just by a small factor, but by more than an order of magnitude on distance and two orders of magnitude on area.

I cannot imagine where he got this data from, but I can assure you that it has no connection with reality.

At the other end of the American political spectrum there are no mass circulation media to present the other extreme, but there are many groups of activists. They—and their friends in Europe—have access to all sorts of mimeographs, xerox machines, and printing presses and they have managed to put out a very large number of pamphlets excoriating the neutron bomb. Like the aim of The Readers' Digest in this matter, the purpose of these pamphlets is not at all to inform, but to persuade and, if possible, to move the reader to action. The total number of readers of such pamphlets is probably smaller than the number that peruse The Readers' Digest, but they may well make up for that quantitative difference in dedication and activism. I will not here refer to any particular pamphlet or organization. Rather, I will simply note one particular notion that has been pushed hard by a great many of them. This is the claim that the objective of those who designed the neutron bomb was to produce a weapon which would "kill people while preserving property." Those who make this claim often go on to add that this evil intention is prima facie evidence of the inhumane motives of the capitalists and other reactionaries behind the whole thing. Now it is true that in a certain special sense the neutron bomb was designed to kill people without destroying property, but the people it is specifically designed to kill are Soviet tank crews and the property that would be left intact is Soviet tanks. Moreover, the reason that it will not destroy Soviet tanks is not that someone in the Pentagon wanted to save them for our own use but simply that the tanks are so tough and sturdy that they are capable of surviving a nearby nuclear explosion.

In asserting that the people on both ideological extremes have got this issue all wrong, I am not in any way denying that the deployment of the neutron bomb poses very troublesome issues, but to understand those issues, we have to begin by getting the facts straight. The neutron bomb is not something wholly new; it is simply another, technologically clever, variant of the basic atomic bomb. It is designed specifically to provide somewhat better control over the killing and destructive effects in certain very special circumstances. Specifically, the neutron bomb, which is intrinsically relatively small in yield (typically about one tenth the power of the Hiroshima bomb) is designed so that for a given lethal effect on crews in tanks or other very hard structures in the immediate target area, it produces somewhat less harmful effects than a standard atomic bomb at middle distances, say, on towns and townspeople a mile or so away. The serious advocates of the neutron bomb argue that as a result of this combination of properties, there will be fewer inhibitions on its use than would be the case of a standard atomic weapon of the same power, and that that in turn—to use the jargon of the military analyst—means that its deployment will provide us with a more "credible deterrent" than would otherwise be the case. To put it the other way around, because an enemy would know that it is easier for us to use the neutron bomb, he will be more thoroughly deterred from taking any action which might lead to its use. Those serious analysts who oppose the neutron bomb agree with the claim that there would be, at least theoretically, fewer inhibitions against its use, but they regard that as a very bad thing and that is precisely why they oppose it. They contend that the higher the "fire break" separating nuclear war from conventional war, the safer and more secure we all will be. As it happens, I am among those opposed to the construction and deployment of the neutron bomb, but the reason is that I favor the abolition of all battlefield nuclear weapons, including the neutron bomb, not because of the peculiar properties of that version of atomic weaponry.

Ethical Considerations

A third major aspect of the overall international security problem is its ethical, or moral, dimension. It seems obvious that ethical considerations must somehow be brought into the process of working out solutions to our policy dilemmas, but it is not at all clear just how to do that.

In practice it often proves uncommonly difficult to combine strategic, political, and ethical ideas into a single package. Let me illustrate this by recounting a particular episode in the history of nuclear weapons.

In August of 1949, the Soviets exploded their first A-bomb. This event produced a major shock in the West, and many proposals for an appropriate American response to it were made. One idea was that the United States should initiate a high priority program to develop and build the hydrogen bomb, a new device then thought of as being 1,000 times as powerful as the standard A-bomb. This idea had first come up early in the wartime Manhattan Project, but it was set aside at the time, and had continued to lay fallow in the early post war years. Along with less draconian suggestions, this H-bomb idea was referred to the General Advisory Committee, established by Congress as part of our governmental apparatus for handling just such questions. The committee was chaired by Robert Oppenheimer and was largely composed of other leaders of the program to develop our own atomic bomb just a few years earlier. The final report of their thinking on this question starts with a discussion of the technical issues, which notes the difficulties and great costs that would be involved, and it expresses some doubts about its military utility, but it concludes by saying there was an even chance that, with an "imaginative and concerted" effort, an H-bomb could be designed and built in five years. The report then turns to what the authors saw as the relevant political and moral issues. The section of the report dealing with these subjects mentions that

> "the extreme dangers to mankind inherent in the proposal wholly outweigh any military advantage," "it represents a threat to the future of the human race that is intolerable," "is necessarily an evil thing considered in any light," "its use...cannot be justified on any ethical ground."

The report concludes:

> "We recommend strongly against" initiating an 'all out' effort "to develop [such] a weapon."

I remember well what followed. Those who opposed the Committee's views—that is, the proponents of the H-bomb—sought and easily found an Achilles heel in the committee's report. In effect, they said that the GAC members had been selected because of their rich scientific knowledge and experience, not because they were outstanding political or moral experts, and yet they had written a report whose final recommendation was based primarily on their moral views. Worse, they claimed, the authors of the report had buttressed this recommendation by adjusting their technical comments to fit their ethical conclusions. This antagonistic view of the report was widely accepted in those circles involved in the issue, and it did, to my personal knowledge, play an important role in the final decision to reject the committee's advice.

This episode is, I think, an example of the real difficulties that commonly arise in trying to meld ethical judgments with practical analyses. Scarcely anyone is a real expert in both of these general areas, and it usually proves easy to repudiate any study that combines such perspectives by focusing on whichever of the two elements is weaker. Such difficulties, I believe, account in part for the long delay—from 1945 to 1975—before very many major religious leaders chose to enter this debate in a very serious way.

In addition to this practical difficulty of mixing ethical thinking with other types of analysis, the ethical ideas—like their political and technical counterparts—are themselves complex and it takes time to work them out. It is not enough to observe that atomic bombs kill people, and to conclude that, since killing is wrong, A-bombs must be banned. Nor is it enough to note that the Soviet state is an "evil, godless empire" which must be resisted by whatever means are available and then to conclude that whatever works against that evil is justified. In recent years, both the Pope and the Archbishop of Canterbury have addressed this problem in public speeches and each has in essence said that military preparedness is justifiable, as is the fighting of a defensive war, but each has also made it clear that the use of nuclear weapons is absolutely not a justifiable means for achieving these ends. At the same time, however, each of these religious leaders has made it clear he understands that the current dreadful situation is intrinsically complex and deeprooted, and neither one has recommended precipitate actions for changing it.

Partly as a response to the yearnings of some of their people, as well as their own deep convictions, the US Catholic bishops have undertaken a long and careful study of the problem. It is clear that they too see the current situation as both urgent and complex. In their "Letter," after noting the need to reverse recent trends and to limit and eventually eliminate our dependency on nuclear weapons, they go on to say:

> "As long as there is hope of this occurring, Catholic moral teaching is willing, while negotiations proceed, to tolerate the possession of nuclear weapons for deterrence as the lesser of the two evils. If that hope were to disappear, the moral attitude of the Catholic Church would certainly have to shift to one of the uncompromising condemnation of both use and possession on such weapons."

That statement is very much more than either just "ban the bomb" or "defense by any means."

It is imperative that these several different ways—that is, political, military and ethical—of looking at the problem be pursued together. We must find some way simultaneously to satisfy both our ethical imperatives and our genuine defense needs.

Zeal Is Not Enough

In finishing, let me summarize by noting that problems as important and urgent as those posed to us by the nuclear arms race do not usually come about because mean spirited persons take deliberately malevolent actions. They commonly arise out of a general situation made worse because well meaning people cannot foresee or forestall all the results of actions which they undertake in the belief that they are all right and necessary.

In such cases, we cannot make the problems go away either by wrapping ourselves in the flag and thinking self-righteous thoughts or by holding hands while standing in a big circle and singing protest songs.

We cannot gain much useful insight into such matters by reading only inspirational literature or polemical tracts.

And we cannot discover how to set things moving in the right direction by attending rallies where we listen to lectures on either Russian political perfidy and military might or on the truly unutterably horrible consequences of nuclear war.

Clearly, zeal is not enough. Anyone wishing to contribute to solving either of our two tightly coupled national security problems must, first, admit that these problems involve deeply rooted and difficult issues, and, second, he or she must make a serious attempt to understand more than just one of their important aspects: political, moral, military or technological.

POWER AND THE
TECHNOLOGICAL
ELITE

Origins of the Lawrence Livermore Laboratory

The nuclear monopoly of the United States ended when the Soviet Union exploded its first atomic bomb on August 29, 1949. After a sharp, brief and secret debate, President Truman determined that the major response to this particular event should be the development of the Superbomb, or H-bomb. This device, estimated then to be 1,000 times as powerful as the first A-bomb, was based on a set of ideas which had been studied on a relatively low priority basis since 1942. In June 1950, North Korea suddenly invaded South Korea, and the United States again found itself at war after less than five years of uncertain peace. In November 1950 great numbers of so-called volunteers from the Chinese People's Liberation Army joined their North Korean comrades and the whole situation took on a more ominous cast.

These historical events as well as three other originally separate strands of developments eventually coalesced and led to the creation of the second U.S. nuclear weapons laboratory at Livermore, California, in the summer of 1952. One of these strands was the determination of Ernest O. Lawrence and Luis Alvarez to involve themselves and their colleagues at Berkeley in some direct, useful and important way in the U.S. response to the first Soviet A-bomb. Another factor was the protracted conflict between Edward Teller, on the one hand, and Norris Bradbury and his senior staff on the other, about how the Los Alamos Laboratory in New Mexico—the first nuclear weapons laboratory in the United States—might best go ahead on the H-bomb program. This conflict finally led Teller to conclude that another laboratory had to be established to do the job adequately. The third factor, a happenstance, was that a small group of Berkeley scientists, including me, had participated in the George shot, the

Figure 1. View of the Livermore weapons lab from the top of its cyclotron building.

first thermonuclear test explosion, of Operation Greenhouse, and thus a small cadre of young men at Berkeley were familiar with the details of thermonuclear weapons design.

On the very day the news of the first Soviet A-bomb became known to the public, Lawrence, Alvarez, Wendell Latimer and others at Berkeley began to ponder the appropriate U.S. response to that event and to search for ways they might participate in such a response. They discussed the matter among themselves, and then traveled to other centers of nuclear research to learn the views of other scientists. Among the places they visited was Los Alamos, where they were particularly interested in learning more about Teller's ideas about the subject.

On the basis of these early explorations, the Berkeley group concluded that they should support Teller's proposals for an urgent, high-priority program at Los Alamos to develop a Superbomb based on the fusion process, and that their group should undertake the design and construction of a reactor which could produce a large excess of neutrons. Teller had explained to them that substantial amounts of tritium—a heavy radioactive form of hydrogen which does not occur naturally—might be needed in the development and manufacture of fusion (hydrogen) bombs, and they knew that the best way to produce tritium was in a reactor specially designed to produce a large excess of neutrons.

The General Advisory Committee (GAC) of the Atomic Energy Commission at its famous meeting in October 1949 decided to recommend against the development of the H-bomb but did agree that the design of such a reactor should be undertaken. The committee also suggested that the program be carried out by the Argonne National Laboratory which had much more relevant experience.

Lawrence and Alvarez were at first disappointed at this turn of events, but they soon responded with an entirely new concept based partly on an idea of Winn Salisbury. It soon became known by its cover name, the Materials Testing Accelerator (MTA), by analogy to the Materials Testing Reactor, which really did have the purpose its name implies. The basic idea of the MTA involved a two step process: first produce large quantities of free neutrons by brute force and, second, absorb these neutrons in suitable materials to produce any of several desired end products—tritium, plutonium, uranium-233 or radiological welfare agents.

To accomplish the first step in this process, they proposed building an enormous particle accelerator 60 feet in diameter and more than 1,000 feet long, capable of producing as much as an ampere of deuterons having energies of several hundreds of millions of volts. Such a device would consume some hundreds of megawatts of energy—about what a medium large reactor produces—and it would, they speculated, produce a somewhat larger number of available free neutrons than that same reactor.

Lawrence first asked Robert Serber and me to make separate theoretical estimates of the production of neutrons in such a device, and then asked me to check them experimentally. I had just received my PhD in physics at Berkeley and had stayed on at Lawrence's laboratory, as what today would be called a postdoctoral fellow. My data revealed that suitably large numbers of neutrons would be produced (5 to 10 neutrons per deuteron at 350 million volts) almost no matter what materials were used to construct the primary target for the deuteron beam.

To accomplish the second step in their novel process, they proposed to surround the primary target with a large secondary target lattice in which the free neutrons produced in the first step would be absorbed in a suitable fertile material. In such a system the uranium that had been depleted of uranium-235—that is, the uranium left over after the isotope separation plants and the plutonium production reactors had largely removed or consumed the uranium-235 it originally contained—could be converted into plutonium. In principle the MTA made it possible to exploit the basic raw material, uranium ore, much more completely and efficiently than would otherwise be the case.

Partly because of these potential benefits and partly because of Lawrence's enthusiasm and his reputation as a scientist who "knew how to get things done," the development and construction of a prototype machine was authorized by Washington. There was, as would be expected, some dispute about such a novel and expensive device, but the Korean war broke out while the project was still in its study phase, and that completely settled the matter for the time being. The Korean war was widely interpreted as a switch by the communists from a program of world conquest by political subversion to a program of direct—if for the moment distant—military action. Since a very large part of our uranium ore came from overseas, mostly the Belgian Congo and South Africa, the worsening world situation was widely perceived as justifying such a Draconian approach to making more complete use of our native supply of ore.

Site Selected

As was customary for such complex nuclear projects, it was decided to proceed by stages. As a first step, a prototype machine suitable for checking out the basic design principles would be built at a site convenient to the Berkeley laboratory. Later, but as soon as possible, a full-scale machine would be built at a site whose selection would be based on other factors, including the availability of sufficient land, operating personnel, electric power, and invulnerability to surprise attack.

A search for a convenient site suitable for the prototype machine resulted in the selection of an abandoned World War II naval air station some 40 miles southeast of Berkeley, at Livermore. The California Research Corporation, a subsidiary of Standard Oil of California, was awarded a contract to build the prototype machine under the scientific guidance of the University of California Radiation Laboratory (UCRL) group, which had by then expanded to include Wolfgang K.H. Panofsky, William Brobeck and many other Berkeley physicists and engineers. It was recognized by all concerned that an industrial contractor would be necessary both to build and to operate a full-scale plant—already planned for construction near St. Louis, Missouri—and the California Research Corporation was brought in at this stage in order for it to gain the necessary experience as quickly as possible.

The Livermore prototype was actually a full-scale model of the front end of the ultimate machine, and an accelerator 60 feet in diameter and 87 feet long was actually built there. It took longer to build than anticipated, and it never did run well. The MTA turned out to involve technological steps too numerous and too large to be taken successfully all at one time.[1] Before all the bugs were worked out of the design it became clear that the main original reason for building it—the threatened shortage of accessible uranium ore—was no longer valid. Within just a few years after the AEC first anticipated the possibility of a serious shortage, it was discovered that simply raising the price of uranium inspired prospectors to discover very large supplies of ore in Colorado and Canada. As a consequence, the political basis of the MTA was eroded and the Livermore machine was shut down, and dismantling started in December 1953. The MTA building and some of the equipment later were used in support of the Sherwood Program, whose purpose was the controlled production of thermonuclear energy.

In the meantime, however, while I was still refining measurements of the potential neutron yield of the MTA, Luis Alvarez approached Hugh Bradner and me in the spring of 1950 to tell us something about the expanding work on the Superbomb at Los Alamos. He said he had recently talked with Edward Teller about the matter, and it appeared that the project could use some assistance from scientific groups at other laboratories. Bradner and I promptly flew to Los Alamos to meet with Teller and others, and we quickly agreed to set up a special group at Berkeley to perform some diagnostic experiments on the first thermonuclear test explosion, code named the George shot of Operation Greenhouse. About 40 persons were members of this special group, code named the Measurements Project, mostly young physicists who we would now call 'postdocs.' In essence, we were to make experimental observations of certain physical phenomena as these unfolded during the first fraction of a microsecond of the thermonuclear explosion.

I recall that several different considerations strongly motivated and inspired me to participate in the hydrogen bomb program. One was my own perception of the growing seriousness of the Cold War, much influenced by my very close personal student-teacher relationship with Lawrence. The Sino-Soviet Bloc had just been formed; Stalin and Mao both said that it was monolithic and that its goal was world revolution. Another inspiration was the scientific and technological challenge of the experiment itself. It was to be the very first occasion in which a thermonuclear reaction took place on the surface of the Earth, and we were to make complex observations extending over a period of less than a millionth of a second.

Five years before, I had played a peripheral role in the Manhattan Project. I had not participated in the Trinity test of the first A-bomb at Alamogordo, and I had only heard about it a week or so after it occurred. This time I was being invited to participate directly in the heart of the matter.

Another strongly favorable consideration was my discovery that Teller, Hans Bethe, Enrico Fermi, John von Neumann, John Wheeler, George Gamow, and others like them were involved in this project at Los Alamos. They were among the greatest men of contemporary science. They were the legendary yet living heroes of young physicists like myself, and I was greatly attracted by the opportunity of working with them and coming to know them personally. Moreover, I was not as yet cleared to see GAC documents of deliberations, so I knew nothing about the arguments opposing the Superbomb except for what I learned secondhand from Teller and Lawrence, who, or course, regarded those arguments as wrong and foolish.*

Most of the preparatory work of our group was done in Berkeley. The pilot set-up of our electronics gear, however, required more room than was readily available at Berkeley, so we used some space in the infirmary at the abandoned naval air station at Livermore. The California Research Corporation was already at work on the MTA project at this site, and its working relationship with the University of California Radiation Laboratory made it natural and simple for it to provide us with all the necessary housekeeping functions.

George Shot

During March and April 1951, most of the members of our special group moved out to Eniwetok Atoll in the Marshall Islands and we set up our equipment in its final form in the shadow of the George device. On May 8, 1951, at Eniwetok Atoll the first thermonuclear text explosion on Earth was successfully conducted. The tritium-deuterium mixture burned well and the various diagnostic experiments, including that of our Berkeley group, were successful in recording the various phenomena that accompanied the explosion.

Some of the members of our Berkeley group, after completing the analysis of their data, participated in the general post-experiment discussions and in some of the future planning sessions. No specific plans for further participation resulted from these discussions, however, so the Berkeley group was disbanded. Its members turned to various other projects, mostly pure research in high energy physics.

Shortly before the George shot, Teller and Stanislaw Ulam had contributed the key ideas that resulted in the invention of the basic Superbomb or H-bomb, a device by which a relatively small fission explosion can induce an arbitrarily large thermonuclear explosion. Contrary to much current folklore, these ideas did not directly involve the notion of using as the thermonuclear fuel the easy-to-handle lithium-deuteride salt rather than

---

* I finally saw the 1949 GAC report for the first time in 1974—a quarter of a century later.

the very awkward liquid deterium; that particular idea goes back to a time several years earlier.

Edward Teller stayed on at Los Alamos for only another six months after the George shot at Eniwetok. The next major experiment, the Mike shot of Operation Ivy,* was to be based on the Teller-Ulam ideas, and Teller participated directly in the determination of its basic configuration. But in November 1951, a year before Mike was to be fired, he left Los Alamos and returned to the University of Chicago. He did so because he felt that the remaining theoretical work needed to get Mike ready could be done just as well without him—Bethe was already scheduled to be at Los Alamos during the final design period—but mainly because of the long standing and increasingly acrimonious argument between him and Bradbury over how to run the laboratory and the hydrogen bomb program.

An Open Secret

In the summer of 1951, only some months after he had come up with the final, capping suggestion in the series of ideas that led to the invention of the Superbomb, Teller had concluded that the establishment of a second, independent laboratory was needed to exploit this new approach in a timely fashion. As Teller later put it:

> It was an open secret, among scientists and government officials, that I did not agree with Norris Bradbury's administration of the thermonuclear program at Los Alamos. Bradbury and I remained friends, but we differed sharply on the most effective ways to produce a hydrogen bomb at the earliest possible date. We even disagreed on the earliest possible date itself, on the timing of our first hydrogen bomb test. The dissension with Bradbury crystallized in my mind the urgent need for more than one nuclear weapons laboratory.[2]

Teller soon succeeded in persuading Gordon Dean, then Chairman of the AEC, to consider the question of establishing a second laboratory. Dean in turn asked the General Advisory Committee of the AEC to review the idea. Except for Willard Libby, like Teller a professor at the University of Chicago and then a new member of the committee, the GAC opposed the idea on the ground that the establishment of a second laboratory would divert talent and resources from Los Alamos and would thus

---

* The first thermonuclear reaction was achieved with the Greenhouse-George test but the Ivy-Mike test on November 1, 1952, achieved the first huge thermonuclear explosion: the detonation measured about 10 megatons and erased the island of Elugelab from the map.

slow down the overall program. Personal elements were probably also importantly involved in the opposition of the GAC to a second laboratory.

Teller's claim that competition was a good thing was often expressed in terms which made it clear he felt the Los Alamos leadership was unimaginative, negative, and otherwise inadequate. It was equally clear that Teller felt much the same way about many of the members of the GAC itself, and so it is not surprising that the GAC supported Los Alamos and Bradbury against what they regarded as an unwarranted personal attack.

Teller also sought support of his ideas in the Air Force. It seemed that they would be the principal user of the hydrogen bomb, and a number of persons at the top of the Air Force very quickly showed great personal and institutional interest in the issues being raised. David Griggs, one of the founders of the Rand Corporation and just then Chief Scientist of the Air Force; James H. Doolittle, a much respected retired General and a high-level consultant to the Air Force; and General Elwood R. Quesada, the commander of the joint task force that had conducted Operation Greenhouse, all became strong partisans of Teller and helped him make further contact with higher Air Force officials.

As a result, Teller and his ideas were warmly received and strongly endorsed by Thomas K. Finletter, Secretary of the Air Force, and his special assistant for R&D, William A.M. Burden.* They, in turn, began to make moves toward establishing a second nuclear weapons laboratory under Air Force sponsorship. In 1951 they actually did arrange to sponsor briefly some nuclear calculations Teller was doing at Chicago. However serious their intentions concerning a full-scale second laboratory may have been, their actions greatly increased the pressure on the AEC either to do something itself or to see its monopoly in the field eliminated.

During this period, Teller also was given the opportunity to brief Secretary of State Dean Acheson, Secretary of Defense Robert Lovett, and Deputy Secretary of Defense William C. Foster. The very fact of these briefings, independent of their specific content or results, put further pressure on the AEC.

In late 1951, Thomas Murray, the Atomic Energy Commissioner most sympathetic to the idea of a second laboratory, got in touch with Ernest Lawrence to discuss the matter with him. Lawrence responded positively and volunteered to study the question further.

Since I had been more deeply involved in the recent thermonuclear program than anyone else at Berkeley, Lawrence in January 1952 asked

---

* Not to be confused with William L. Borden, then Executive Director of the staff of the Joint Committee on Atomic Energy, and the man who later first formally charged J. Robert Oppenheimer with being a Soviet agent.

me what I thought. As a direct result of Lawrence's inquiry, I made a series of extended trips to Los Alamos, Chicago, and Washington, where I discussed the matter with Teller and most of the people named above plus a few others, including Army General Kenneth Fields, then the AEC's director of Military Applications, and his deputy, Navy Captain John T. Hayward. I found the whole affair heady and exciting, and I was readily persuaded to Teller's point of view. I reported to Lawrence that I, too, felt it would probably be useful to establish a second laboratory.

The idea of doing so at the Livermore site was, for us, a natural one, and we suggested it immediately to the AEC. This proposal to establish facilities at Livermore as a branch of the University of California Radiation Laboratory instead of simply establishing one somewhere under an unspecified aegis, changed the nature of the argument. It clearly meant much less initial expense and an immediate, if small, cadre of people ready to go to work right away. As a result, as GAC Chairman Oppenheimer later recalled, the GAC and the AEC "approved the second laboratory as now conceived because there was an existing installation, and it could be done gradually and without harm to Los Alamos."[3]

As I recall it, Lawrence and Teller felt at the time that Oppenheimer himself was still really opposed to a second laboratory but that under the new circumstances he had no other choice. Even so, during that year I met with Oppenheimer at Princeton and discussed the plans for the Livermore laboratory. He received me in a personally friendly fashion, but I cannot recall his being of any particular help.

The precise nature of the plans for the new laboratory, however, primarily reflected Lawrence's ideas about how to go about such things and deviated considerably from Teller's views of what should be done. In essence, Lawrence firmly believed that if a group of bright young men were simply sent off in the right direction with a reasonable level of support, they would end up in the right place. He did not believe that the goals needed to be spelled out in great detail or that it was necessary that the leadership consist of persons who were already well known. Teller, on the other hand, had become deeply suspicious of the intentions of the AEC leadership, and he wanted something more nearly analogous to the 1943 plans for Los Alamos—a plan for a laboratory with a well-defined goal that would be led by a large cadre of famous scientists.

To complicate matters, during that spring Lawrence, suffering from a chronic illness, spent much time away from Berkeley on long rest trips. As a result I was left pretty much on my own to draw up the specific plans for a second laboratory with nothing except the most general guidance from my immediate superior. However, as a result of 10 years close association, I both clearly understood and firmly agreed with Lawrence's approach to 'big science,' and I generated plans which he always warmly endorsed when he had a chance to review them.

Finally, and in close accordance with Lawrence's (and my) views of the matter, the AEC in June 1952 approved the establishment of a branch of the Berkeley laboratory at Livermore to assist in the thermonuclear weapons program by conducting diagnostic experiments during weapons tests and other related research. The question of how soon, or even whether, the Livermore Laboratory would actually engage directly in weapons development was left open, however. The AEC's official planning document described the mission of the Livermore Laboratory this way:

> a. Development and experimentation in methods and equipment for securing diagnostic information on behavior of thermonuclear devices and the conduct of such instrumentation programs in support of tests of thermonuclear devices, in close collaboration with the Los Alamos Scientific Laboratory.

> b. While the work authorized above is the immediate objective of this proposal, the Commission hopes that the group at UCRL (Livermore) *will eventually suggest broader programs of thermonuclear research* to be carried out by UCRL or elsewhere. (Emphasis added.)[4]

Lawrence felt that this statement of intentions provided an adequate base upon which to build a second weapons laboratory. I would have preferred something more concrete, but I was prepared to accept it as a place to start from. Teller, on the other hand, found the vagueness of the AEC's plans for the Livermore Laboratory entirely unsatisfactory.

As a result, in early July he told Lawrence, Gordon Dean, myself and others that he would have nothing further to do with the plans for establishing a laboratory at Livermore. The Berkeley administration was prepared to go ahead anyway. However, at the insistence of Captain John T. Hayward (the Deputy Director of the AEC's Division of Military Applications), more than anyone else, intense negotiations were resumed among all concerned. Within days this led to a firm commitment on the part of Gordon Dean that thermonuclear weapons development would be included in the Livermore program from the outset, and also to a renewed commitment on the part of Teller to join the laboratory.

The Livermore Laboratory was launched in September 1952. I became the director, and the Scientific Steering Committee included Teller, Harold Brown, John S. Foster, Jr., Arthur T. Biehl, and a few others. Teller, because of his obvious special status, was given veto authority over the decisions of the committee but otherwise had no formal authority. Brown was put in charge of the development of thermonuclear weapons at Livermore; Biehl at first and then Foster was put in charge of the development of improved fission weapons. There were some early problems in the administration of the theoretical division; but these were resolved, on a temporary basis, by appointing Richard Latter as acting head of the division. Latter, a Rand physicist, was temporarily on loan four days a week to the Livermore Laboratory. After about a year, he was replaced by Mark M. Mills, who remained in that position until his death in a helicopter crash at Eniwetok in early 1958. These organizational arrangements, although they contained some peculiar elements, worked out very well and none of the strained relationships that had surrounded Teller at Los Alamos developed at Livermore.

L'Affaire Mike

The relationship between the Livermore and Los Alamos laboratories was strained from the beginning, and rapidly grew worse. One of the causes of this state of affairs involved the question of who deserved the credit for Mike, the first large thermonuclear device.

In November 1952, two months after the Livermore Laboratory was established, Mike was exploded during Operation Ivy. Based on ideas by Teller, Ulam, and many others, it was designed, built and tested by the Los Alamos Laboratory. Los Alamos was aided in those tasks by a number of other institutions, including the Naval Research Laboratory and the Matterhorn-B group at Princeton; but not by the Livermore Laboratory, which then was completely absorbed in simply coming into being. Nevertheless, the press commonly gave Teller, and by implication the Livermore Laboratory, the whole credit.

The reasons for this error were Teller's presence at Livermore and the absurdly strict secrecy policy on the part of the AEC. The AEC, purportedly for security reasons, refused either to make or to allow any comment whatsoever on the origin and nature of Mike, not even the simple denial of its being created by Livermore. This situation continued for almost two years. It was not until after Shepley and Blair brought out a biased account of the matter in 1954 and excerpts were published in *Life* magazine that Bradbury was finally allowed to respond by holding a press conference giving his version of history.[5]

Relations between Livermore and Los Alamos, already bad, became worse during the hearings on J. Robert Oppenheimer's security clearance. Oppenheimer's clearance had been lifted in late 1953 and hearings in the matter held in the spring of 1954. Despite elaborate efforts to assert the contrary, the real stimulus for the removal of Oppenheimer's clearance was his opposition to the program for the development of the H-bomb.

Teller, along with Lawrence, Alvarez, Wendell Latimer, and some others at the University of California provided some of the most important arguments in support of the government's case. On the other hand, both the veterans of the wartime Los Alamos Laboratory as well as its leadership during that time strongly supported Oppenheimer. All during those very trying times, the Los Alamos administration treated the Livermore leadership with formal correctness and provided some much needed technological assistance to the new laboratory. But some leading scientists at Los Alamos made it clear that Teller was personally unwelcome there. Since Teller was such a central figure at Livermore, this feeling inevitably affected the relationship between the laboratories.

On top of all that, the first Livermore tests—in Nevada in 1953 and at Bikini in 1954—went poorly. Given the attacks on the quality of the leadership at Los Alamos that were part of the arguments supporting the establishment of a second laboratory, it is not surprising that some Los Alamos scientists filled the air with horse laughs on those occasions.

The Livermore Laboratory flourished nonetheless and eventually produced its share of new weapons ideas and designs. From its first year of operations (fiscal 1953) with a budget of $3.5 million and a staff of 698 the laboratory grew steadily. Five years later (fiscal year 1958) during my last year as director, the staff passed the 3,000 mark and the budget reached $55 million. In another five years the staff passed 5,000 and the budget reached $127 million. Shortly after that, the laboratory stabilized at a population of about 5,500.

The establishment of the second nuclear weapons laboratory at Livermore—renamed the Lawrence Livermore Laboratory after Lawrence's death in 1958—in effect doubled the number of scientists, engineers and technicians working on nuclear weapons. But this doubling of the number of persons involved did much more than simply increase the rate of progress by a factor of two. As Teller said in explaining why he left Los Alamos and undertook the promotion of a second laboratory, technology, like so many other elements of human endeavor, does thrive on competition. I have little doubt that the Livermore Laboratory was seen as the competition by Los Alamos, and I know for a fact that we at Livermore saw ourselves as being in competition with them.

Within 10 years the new burst of intellectual energy, which both laboratories focused on nuclear weapons design problems, resulted not only in a great profusion of relatively large hydrogen bombs for ICBMs and the like, but also in varieties of nuclear weapons that weighed less than 100 pounds (as the Davy Crockett), that were only inches in diameter (as for artillery shells), and that were on the average sufficiently efficient in their use of nuclear materials so that they could number many tens of thousands.

A Great Leap Forward?

The introduction of such a large variety of weapons led directly to their being stored in a great many different locations abroad as well as at home, and to a correspondingly great proliferation in the number of military command elements having immediate control of the individual weapons. There can be little doubt that the great leap forward in nuclear weapons design and deployment that occurred in the 1950s resulted in part from the creation of the Lawrence Livermore Laboratory. Whether this accelerated rate of progress has had a beneficial or harmful effect on national security is another interesting and important question. I believe though the net effect has been harmful.

My purpose in recounting the history of the Livermore Laboratory is twofold. One purpose is to help today's citizens understand a little better how the world they see about them came to be. Another purpose is to help provide a piece of the factual background needed for the current restudy of America's national goals and needs.

A third of a century ago, the terrible events that made up the early stages of World War II brought about the creation of the U.S. nuclear weapons development program. A quarter century ago, the very threatening events that marked the darkest years of the Cold War stimulated a large increase in the pace of the U.S. nuclear program, including but not limited to the establishment of the second nuclear weapons laboratory. Now, with our longest war ended and peace reestablished, it is necessary and appropriate to undertake once again the same kind of questioning and replanning of the nuclear activities of the United States that took place in 1949–1952. Surely reasons or rationalizations a third and a quarter of a century old cannot be automatically accepted as the basis for current policies. Contemporary programs must be based on contemporary reasons.

References

1. The only published account of the MTA project is a 'gee whiz' article by Allen P. Armagnac, "The Most Fantastic Atom-Smasher," Popular Science, Nov. 1958, pp. 108ff.
2. Edward Teller with Allen Brown, The Legacy of Hiroshima (Garden City, N.Y.: Doubleday & Co., 1962), p. 54-55.
3. In the matter of J. Robert Oppenheimer (Cambridge, Mass.: MIT Press, 1971), p. 248.
4. Atomic Energy Commission, Director of Military Applications. Thermonuclear Research at the University of California Radiation Laboratory, AEC 425/20 (Washington, D.C.: The Commission, June 13, 1952).
5. Shepley and Blair, The Hydrogen Bomb, New York, 1954. Teller's version is given in The Legacy of Hiroshima, and in Science, Feb. 25, 1955, p. 268.

Debate Over the Hydrogen Bomb

In 1948 Czechoslovak Communists carried out a coup in the shadow of the Red Army and replaced the government of that country with one subservient to Moscow. Also in 1948 the Russians unsuccessfully attempted to force the Western allies out of Berlin by blockading all land transport routes to the city. In early 1949 the Communist People's Liberation Army captured Peking and soon afterward established the People's Republic of China. Taken together, these and similar but less dramatic events were generally perceived in the West as resulting in the creation of a monolithic and aggressive alliance stretching the full length of the Eurasian continent, encompassing almost half of the world's people and threatening much of the rest. Then in the fall of 1949 the Russians exploded their first atomic bomb and ended the brief American nuclear monopoly.

At the end of World War II most atomic scientists in the U.S. had estimated that the U.S.S.R. would need four or five years to make a bomb based on the nuclear-fission principle; the time interval from the first American test to the first Russian one turned out to be four years and six weeks. Even so, nearly everyone, including most U.S. Government officials and most members of Congress, reacted to the event as if it were a great surprise. Many of them had either forgotten or had never known the experts' original estimates, and in any case the accomplishment simply did not fit the almost universal view of the U.S.S.R. as a technologically backward nation.

Besides being a great surprise the Russian test explosion was a singularly unpleasant one. The U.S. nuclear monopoly had been seen by many as compensating for the difference between the hordes of conscripts supposedly available to the Communist bloc and the smaller armies available to the Western countries. Coming as it did at a time when virtually all Americans saw the cold war as rapidly going from bad to worse, the Russian test was seen as a challenge that demanded a reply. The immediate challenge being nuclear, a particularly intensive search for an appropriate response was conducted by those responsible for U.S. nuclear policy.

Most of the proposed responses involved substantial but evolutionary changes in the current U.S. nuclear programs: expand the search for additional supplies of fissionable material, step up the production of atomic weapons, adapt such weapons to a broader range of delivery vehicles and end uses, and the like. One proposal was radically different. It called for the fastest possible development of the hydrogen bomb, which was widely referred to at the time as the superbomb (or simply the Super). This weapon, based on the entirely new and as yet untested principle of thermonuclear fusion, was estimated to have the potential of being 1,000 or more times as powerful as the fission bombs that had marked the end of World War II. Work on the theory of the superbomb had already been going on for seven years, but it had never had a very high priority, and so far it had yielded no practical result. A number of scientists and politicians endorsed the proposal, but for years Edward Teller had been its leading advocate. The superbomb proposal led to a brief, intense and highly secret debate.

The opponents of the proposal argued that neither the possession of the new bomb nor the initiation of its development was necessary for maintaining the national security of the U.S., and that under such circumstances it would be morally wrong to initiate the development of such an enormously powerful and destructive weapon. In essence they contended that the world ought to avoid the development and stockpiling of the superbomb if it was at all possible, and that a U.S. decision to forgo it was a necessary precondition for persuading others to do likewise. Furthermore, they concluded that the dynamism and relative status of U.S. nuclear technology were such that the U.S. could safely run the risk that the U.S.S.R. might not practice similar restraint and would instead initiate a secret program of its own.

The advocates of the superbomb maintained that the successful achievement of such a bomb by the Russians was only a matter of time, and so at best our forgoing it would amount to a deliberate decision to become a second-class power, and at worst it would be equivalent to surrender. They added that undertaking the development of the superbomb was morally no different from developing any other weapon.

The secret debate about what the American response ought to be took place within the government itself. Many organizations were involved, including the National Security Council, the Department of Defense, the Department of State and the Congressional Joint Committee on Atomic Energy, but the initial focus of the debate lay within the Atomic Energy Commission.

The early official reaction of the AEC's Los Alamos Scientific Laboratory to the Russian test was a proposal to step up the pace of the nuclear-weapons program in all areas. Among other measures, Norris E. Bradbury, the director, recommended that the laboratory go on a six-day work week and that they expand the staff, particularly in theoretical physics.

This acceleration was to include not only programs for improving fission weapons by conventional means but also tests of the booster principle. (In this context "booster" refers to a synergistic process in which the explosion of a comparatively large mass of fissionable fuel, say plutonium or uranium 235, causes a comparatively small mass of thermonuclear fuel, say deuterium and tritium, to burn violently. The high-energy neutrons produced in the thermonuclear process then react back on the fission explosion, boosting, or accelerating, it to a higher efficiency than would otherwise be the case.) The booster concept had been known for several years, and even before the Russian test it had been agreed to include a full-scale experimental test of the process in a 1951 nuclear-test series. The AEC's Director of Military Application, General James McCormack, Jr., received these proposals from the Los Alamos laboratory and sought the advice of the AEC's scientific experts on them. Other AEC division heads were similarly studying proposals for expanding the relevant programs within their jurisdiction.

At about the same time, Teller, then at Los Alamos, Ernest O. Lawrence, Luis W. Alvarez and Wendell M. Latimer at the University of California at Berkeley, Robert LeBaron at the Department of Defense, Senator Brien McMahon, Chairman of the Joint Committee on Atomic Energy, his staff chief William L. Borden and Commissioner Lewis L. Strauss of the AEC had all come to focus on the superbomb as the main element of the answer to the Russian atomic bomb, and they initiated a concerted effort to bring the entire Government around to their point of view as quickly as possible.

As a result of all this concern and activity the AEC called for a special meeting of its General Advisory Committee to be held as soon as possible. This committee was one of the special mechanisms established by the Atomic Energy Act of 1946 for the purpose of managing the postwar development of nuclear energy in the U.S. Its function was to provide the AEC with scientific and technical advice concerning its programs. The members of the committee were all men who had been scientific or technological leaders in major wartime projects. J. Robert Oppenheimer, who was elected chairman of the committee, had been director of the Los Alamos laboratory during the period when the first atomic bomb had been designed and built there. The other members, all scientists, were Oliver E.

Buckley, James B. Conant, Lee A. DuBridge, Enrico Fermi, I.I. Rabi, Hartley Rowe, Glenn T. Seaborg and Cyril S. Smith. Many of the members of this committee and later General Advisory committees also served on other high-level standing committees and some key ad hoc committees, and so a rather complex web of interlocking advisory-committee memberships developed. As a result several of these men, including Oppenheimer, had much more influence than the simple sum of their various committee memberships would indicate.

Oppenheimer was not only the formal leader of the General Advisory Committee but also, by virtue of his personality and background, its natural leader. His views were therefore of special importance in setting the tone and determining the content of the committee's reports in this matter, as in most other matters.

Throughout Oppenheimer's service on the committee he generally supported the various programs designed to produce and improve nuclear weapons. At the same time he was deeply troubled by what he had wrought at Los Alamos, and he found the notion of bombs of unlimited power particularly repugnant. Ever since the end of the war he had devoted much of his attention to promoting the international control of atomic energy with the ultimate objective of achieving nuclear disarmament. He and Rabi had in effect been the originators of the plan for nuclear-arms control that later became known as the Baruch Plan. Oppenheimer's inner feelings about nuclear weapons were clearly revealed in an often quoted remark: "In some sort of crude sense which no vulgarity, no humor, no overstatement can quite extinguish, the physicists have known sin, and this is a knowledge which they cannot lose."

The call for the special meeting, in addition to raising the question of a high-priority program to develop the Super, also asked the committee to consider priorities in the broadest sense, including "whether the Commission is now doing things we ought to do to serve the paramount objectives of the common defense and security." As for the Super, the Commission wanted to know "whether the nation would use such a weapon if it could be built and what its military worth would be in relation to fission weapons." The meeting of the Oppenheimer committee was held on October 29 and 30, 1949; all members were present except Seaborg, who was in Europe. The committee in the course of its deliberations heard from many outside experts in various relevant fields, including George F. Kennan, the noted student of Russian affairs, General Omar Bradley, Chairman of the Joint Chiefs of Staff, and the physicists H.A. Bethe and Robert Serber. Toward the end of the two-day meeting the advisers had a long session with the Atomic Energy commissioners and with their intelligence staff.

The next day the committee prepared its report.

The General Advisory Committee report consisted of three separate sections that were unanimously agreed on and two addenda giving certain specific minority views. In 1974 the report was almost entirely declassified, with only a very few purely technical details remaining secret.

Part I of the report dealt with all pertinent questions other than those directly involving the Super. The advisory committee in effect reacted favorably to the proposals of the various AEC division directors with regard to the expansion of the facilities for separating uranium isotopes, for producing plutonium and for increasing the supplies of uranium ore. These proposals and the committee's endorsement of them were followed eventually by a substantial increase in the rate of production of fissionable materials.

In Part I the committee also recommended the acceleration of research and development work on fission bombs, particularly for tactical purposes. Under the heading "Tactical Delivery" the report stated: "The General Advisory Committee recommends to the Commission an intensification of efforts to make atomic weapons available for tactical purposes, and to give attention to the problem of integration of bomb and carrier design in this field."

This quoted paragraph deserves special emphasis, since it has often been suggested that Oppenheimer, Conant and some of the others opposed nuclear weapons in general. They did apparently find them all repugnant, and they did try hard to create an international control organization that would ultimately lead to their universal abolition. In the absence of any international arms-limitation agreements with reliable control mechanisms, however, they explicitly recognized the need to possess nuclear weapons, particularly for tactical and defensive purposes, and they regularly promoted programs designed to increase their variety, flexibility, efficiency and numbers. For the next few years, right up to the time Oppenheimer's security clearance was removed, he continued strongly to promote the idea of an expanded arsenal of tactical nuclear weapons. The only type of nuclear weapon the General Advisory Committee opposed— and it did so openly—was the Super.

Part I of the report further recommended that a project be initiated for the purpose of producing "freely absorbable neutrons" to be used for the production of uranium 233, tritium and other potentially useful nuclear materials. Perhaps most important of all in the present context, Part I also stated: "We strongly favor, subject to favorable outcome of the 1951 Eniwetok tests, the booster program." This short phrase makes it abundantly clear that the Oppenheimer committee favored conducting research fundamental to understanding the thermonuclear process, and that its grave reservations were specifically and solely focused on one particular application of the fusion process.

Part II discussed the Super. It outlined what was known about the hydrogen bomb, and it expanded on the unusual difficulties its development presented, but it concluded that the bomb could probably be built. In part it said: "It is notable that there appears to be no experimental approach short of actual test which will substantially add to our conviction that a given model will or will not work. Thus, we are faced with a development which cannot be carried to the point of conviction without the actual construction and demonstration of the essential elements of the weapon in question. A final point that needs to be stressed is that many tests may be required before a workable model has been evolved or before it has been established beyond reasonable doubt that no such model can be evolved. Although we are not able to give a specific probability rating for any given model, *we believe that an imaginative and concerted attack on the problem has a better than even chance of producing the weapon within five years.*"

That last sentence (the italics are added) deserves special emphasis. It has been suggested in the past that the General Advisory Committee in general and Oppenheimer in particular were deceptive in their analysis of the technological prospects of the Super; in other words, that they deliberately painted a falsely gloomy picture of its possibilities in order to reinforce their basically ethical opposition to its development. Given the technological circumstances then prevailing, this statement of the program's prospects could hardly have been more positive.

The report then discussed what might be called the "strategic economics" of the Super as they were then conceived: "A second characteristic of the super bomb is that once the problem of initiation has been solved, there is no limit to the explosive power of the bomb itself except that imposed by requirements of delivery. [In addition there will be] very grave contamination problems which can easily be made more acute, and may possibly be rendered less acute, by surrounding the deuterium with uranium or other material....It is clearly impossible with the vagueness of design and the uncertainty as to performance as we have them at present to give anything like a cost estimate of the super. If one uses the strict criteria of damage area per dollar, it appears uncertain to us whether the super will be cheaper or more expensive than the fission bombs." In Part III the committee members got to what to them was the heart of the matter, the question of whether or not the Super should be developed: "Although the members of the Advisory Committee are not unanimous in their proposals as to what should be done with regard to the super bomb, there are certain elements of unanimity among us. We all hope that by one means or another the development of these weapons can be avoided. We are all reluctant to see the United States take the initiative in precipitating this development. We are all agreed that it would be wrong at the present moment to commit ourselves to an all-out effort toward its development.

"We are somewhat divided as to the nature of the commitment not to develop the weapon. The majority feel that this should be an unqualified commitment. Others feel that it should be made conditional on the response of the Soviet government to a proposal to renounce such development. The Committee recommends that enough be declassified about the super bomb so that a public statement of policy can be made at this time."

In the two addenda those members of the committee who were present (that is, all except Seaborg) explained their reasons for their proposed "commitment not to develop the weapon." The first addendum was written by Conant and signed by Rowe, Smith, DuBridge, Buckley and Oppenheimer. In part it said: "We base our recommendation on our belief that the extreme dangers to mankind inherent in the proposal wholly outweigh any military advantage that could come from this development. Let it be clearly realized that this is a super weapon; it is in a totally different category from an atomic bomb. The reason for developing such super bombs would be to have the capacity to devastate a vast area with a single bomb. Its use would involve a decision to slaughter a vast number of civilians. We are alarmed as to the possible global effects of the radioactivity generated by the explosion of a few super bombs of conceivable magnitude. If super bombs will work at all, there is no inherent limit in the destructive power that may be attained with them. Therefore, a super bomb might become a weapon of genocide.

"We believe a super bomb should never be produced. Mankind would be far better off not to have a demonstration of the feasibility of such a weapon until the present climate of world opinion changes.

"In determining not to proceed to develop the super bomb, we see a unique opportunity of providing by example some limitations on the totality of war and thus of limiting the fear and arousing the hopes of mankind."

Contrary to a frequently suggested notion, the members of the Oppenheimer committee were not at all unmindful of the possibility that the U.S.S.R. might develop the Super no matter what the U.S. did. Indeed, they regarded it as entirely possible and explained why it would not be crucial: "To the argument that the Russians may succeed in developing this weapon, we would reply that our undertaking it will not prove a deterrent to them. Should they use the weapon against us, reprisals by our large stock of atomic bombs would be comparably effective to the use of a 'Super.'"

The minority addendum, signed by Fermi and Rabi, expressed even stronger opposition to the Super but loosely coupled an American renunciation with a proposal for a worldwide pledge not to proceed: "It is clear that the use of such a weapon cannot be justified on any ethical ground which gives a human being a certain individuality and dignity even if he happens to be a resident of an enemy country.

"The fact that no limits exist to the destructiveness of this weapon makes its very existence and the knowledge of its construction a danger to humanity as a whole. It is necessarily an evil thing considered in any light.

"For these reasons we believe it important for the President of the United States to tell the American public, and the world, that we think it wrong on fundamental ethical principles to initiate a program of development of such a weapon. At the same time it would be appropriate to invite the nations of the world to join in a solemn pledge not to proceed in the development of construction of weapons of this category."

As with the majority, Fermi and Rabi also explicitly took up the possibility that the Russians might proceed on their own, or even go back on a pledge not to: "If such a pledge were accepted even without control machinery, it appears highly probably that an advanced state of development leading to a test by another power could be detected by available physical means. Furthermore, we have in our possession, in our stockpile of atomic bombs, the means for adequate 'military' retaliation for the production or use of a 'Super.'"

On December 2 and 3, five weeks after the special meeting, the General Advisory Committee convened for one of its regularly scheduled meetings and carefully reviewed the question of the Super once again. According to Richard G. Hewlett, the AEC's official historian, Oppenheimer reported to the commissioners that no member wished to change the views expressed in the October 30 report.

For a time it appeared that the views of the Oppenheimer committee had a chance of being accepted. David E. Lilienthal, chairman of the AEC, was receptive to the committee's point of view. He similarly favored two parallel responses to the Russian test: (1) increasing the production of fission weapons and developing a greater variety of them, particularly for tactical situations, and (2) officially announcing our intention to refrain from proceeding with the Super while simultaneously reopening and intensifying the search for international control of all kinds of weapons of mass destruction. Lilienthal considered the complete reliance on

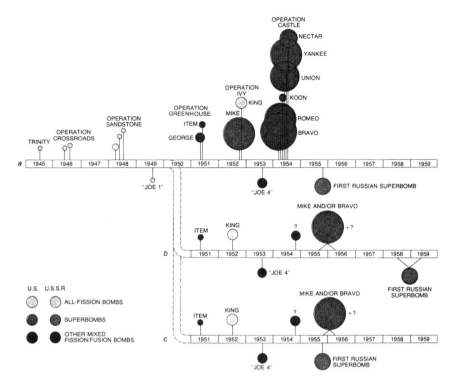

Figure 1. Two hypothetical outcomes are postulated in an effort to evaluate how much risk would have been involved in a U.S. decision not to proceed with the superbomb. They are depicted in this historical chart as branches of the time line representing the actual world (a). The first branch is referred to by the author as the "most probable alterative world" (b), the second as the "worst plausible alternative world" (c). Both branches originate at January, 1950, the date President Truman announced his decision to go ahead with the superbomb. The circles denote nuclear-test explosions; the labels are U.S. code names. Area of each circle is proportional to the region that could be destroyed by the bomb. Bombs of "nominal" size (less than 50 kilotons) have been omitted after 1950.

weapons of mass destruction to be a fundamental weakness in U.S. policy, and he viewed a "crash" program on the hydrogen bomb as foreclosing what might be the last good opportunity to base U.S. foreign policy on "something better than a headlong rush into war with weapons of mass destruction." "We are," he said, "today relying on an asset that is readily depreciating for us, i.e., weapons of mass destruction. [A decision to go ahead with the Super] would tend to confuse and, unwittingly, hide that fact and make it more difficult to find some other course."

As we know now, the advice of the Oppenheimer committee was rejected. Early in 1950 President Truman, acting on the basis of his own political judgment and on the totality of the advice he had received on the matter, issued directives designed to set in motion a major U.S. program to develop the hydrogen bomb.

It is not possible here to give a full description of what happened next, but the following chronological outline of the Russian and American superbomb programs is designed to show how the "race" for the superbomb did in fact come out, and to facilitate making judgments about the General Advisory Committee's advice and about "what might have been."

First of all, it is now known that both countries initiated high-priority programs for the development of a hydrogen bomb at about the same time (late 1949–early 1950), and both had been seriously studying the subject for some years before that.

The first U.S. test series that included experiments designed to investigate thermonuclear explosions took place at Eniwetok in the spring of 1951. Known as Operation Greenhouse, the series included two thermonuclear experiments. One, with the code name Item, was a test of the booster principle. This experiment, it must be emphasized, was planned and programmed before the first Russian atomic-bomb test. The other (which actually took place first) was called George. It was a response to Joe 1, as the first Russian atomic-bomb test was called by the U.S. intelligence establishment. Reduced to its essentials, the purpose of the experiment was to show, as a minimum, that a thermonuclear reaction could under ideal conditions be made to proceed in an experimental device. This experiment came to play a key role in the Super program. As Teller later put it: "We needed a significant test. Without such a test no one of us could have had the confidence to proceed further along speculations, inventions and the difficult choice of the most promising possibility. This test was to play the role of a pilot plant in our development."

The George shot served its purpose well. During the final stages of calculations concerned with the expected performance of this device, Teller and Stanislaw Ulam came up with the climactic idea that made it possible to achieve the goal of the superbomb program: they invented a configuration that would make it possible for a small fission explosion to ignite an arbitrarily large fusion explosion.

The first test of a device designed to ignite a large thermonuclear explosion by means of a comparatively small quantity of fissionable material took place at Eniwetok on November 1, 1952 (local time). The device, known as Mike, produced a tremendous explosion, equivalent in its energy release to 10 megatons (10 million tons) of TNT. As had been repeatedly predicted since the early 1940's, the yield was roughly 1,000 times larger than the yield of the first atomic bombs. For certain practical reasons relating to the pioneering nature of the test, this first version of the Teller-Ulam configuration had liquid deuterium as its thermonuclear fuel. (The last point needs special emphasis. The Teller-Ulam invention, contrary to folklore, was not the notion of substituting easy-to-handle lithium deuteride for the hard-to-handle liquid deuterium. That possibility had been recognized several years earlier.)

Also in November, 1952, the U.S. tested a very powerful fission bomb, with the code name King, that had an explosive yield of 500 kilotons, or half a megaton. Its purpose was to provide the U.S. with an extraordinarily powerful bomb by means of a straightforward extension of fission-weapons technology, in case such large bombs should become necessary for any strategic or political reason. Originally proposed by Bethe as a substitute for the Super program, it became instead a backup for it.

The first Russian explosion involving fusion reactions took place on August 12, 1953. Russian descriptions of this test and later ones confirm that it was not a superbomb. It was only some tens of times as big as the standard atomic bombs of the day, about the same size as but probably smaller than King, the largest U.S. fission bomb. It evidently involved one of several possible straightforward configurations for igniting a fairly small amount of fissionable material. It was the first device anywhere to use lithium deuteride as a fuel, and presumably it could have been readily converted into a practical weapon if there had been any point in doing so. It seems to have been a development step the U.S. bypassed in its successful search for a configuration that would make it possible to produce an arbitrarily large explosion with a relatively small quantity of fissionable material.

In the spring of 1954 the U.S. successfully exploded six more variants of the superbomb in Operation Castle. Their yields varied widely. The first and most famous of these tests, with the code name Bravo, was exploded on March 1, 1954, at Bikini. Its design, which was initiated before the Mike explosion, also incorporated the Teller-Ulam configuration, but it had the more practical lithium deuteride as its thermonuclear fuel. Bravo's yield was 15 megatons, even more than Mike's, and it was readily adaptable to delivery by aircraft.

On November 23, 1955, the U.S.S.R. exploded a bomb that had a yield of a few megatons. According to a statement made by Secretary Khrushchev, this device involved an "important new achievement" that made it possible by "using a relatively small quantity of fissionable material...to produce an explosion of several megatons." Khrushchev's remark is generally taken as confirmation that the test was the first one in which the Russians incorporated the Teller-Ulam configuration or something like it. It also used lithium deuteride as a fuel and was therefore a true superbomb, comparable to the U.S. Bravo device exploded 20 months earlier, except for its yield, which was still probably only about a fifth the yield of Bravo.

With this chronology in mind, what can one say about what might have happened if the U.S. had followed the advice of Oppenheimer and the rest of the General Advisory Committee, backed by Lilienthal and the majority of the AEC commissioners, and had not initiated a program for the specific purpose of developing the Super in the spring of 1950?

At best the invention of very large, comparatively inexpensive bombs of the Super type would have been forestalled or substantially delayed. Very probably the work on the booster principle, which presumably would still have gone forward, would have led eventually to the ideas underlying the design of very big bombs, but those ideas might well have been delayed until both President Eisenhower and Secretary Khrushchev were in power. Those two leaders were both more seriously interested in arms-limitation agreements than their predecessors had been, and it is at least possible that they might have been able to deal successfully with the superbomb. To be sure, such a favorable result was not very probable (certainly it had much less than an even chance of coming about), but its achievement would have been so beneficial to mankind that at least some small risk was clearly worth running.

To evaluate just how much risk would have been involved let us next examine three other outcomes, which I have labeled the "actual world," "the most probable alternative world," and the "worst plausible alternative world."

In both of the hypothetical alternative worlds I assume that the U.S. would have forgone the development of the Super but that the Russians would have ignored this American restraint and would have proceeded at first just as they did in the actual world. I also assume that the U.S would have vigorously followed the positive elements of the Oppenheimer committee's advice; thus the booster project and other ideas for improving fission bombs would have been accelerated. The difference between the most probable alternative world and the worst plausible alternative world lies in the timing of the test of the first Russian superbomb. In the worst plausible world I assume that this test would have come on the same date that it did in the actual world. In the most probable alternative world, however, I assume that the test would have been substantially delayed.

In both of the two hypothetical alternative worlds, then, the Russians in August, 1953, would have exploded Joe 4, a large bomb deriving part of its explosive energy from a thermonuclear fuel and yielding a few hundred kilotons. Such a device, however, would have had no real effect on the "balance of terror." In both alternative worlds the U.S would surely have already tested the 500-kiloton all-fission bomb in November, 1952 (or probably earlier, since the timing of Operation Ivy was determined by the availability of the much more complicated Mike device). Therefore the explosion of Joe 4 would have meant that the U.S.S.R. had caught up with but not surpassed the U.S. insofar as the capability of producing enormous damage in a single explosion was concerned.

Then what would have happened? From that point the Russians might conceivably still have gone on to produce their multimegaton explosion in November, 1955, but I think it is very probable that they would not have done so until much later. In the actual world they had the powerful stimulus of knowing from our November 1952 test that there was some much better, probably novel way of designing hydrogen bombs so as to produce much larger explosions than the one they demonstrated in their August 1953 experiment. A careful analysis of the radioactive fallout from the Mike explosion may well have provided them with useful information concerning how to go about it. In the hypothetical world where the U.S. would have followed the Oppenheimer-Lilienthal advice that stimulus and information would have been absent. Moreover, a comparison of the way nuclear-weapons technology advanced in the U.S. and the U.S.S.R. during that period makes it seem likely there would have been a much longer delay—probably some years—before they took that big and novel a step without such stimuli and information. Therefore in the most probable alternative world the first Russian superbomb test would have been delayed until well after the first American superbomb test (in other words, delayed until 1957 or 1958), whereas in the worst plausible alternative world it would have occurred just when it did in the actual world: in August, 1955.

What would the U.S. have done in the meantime?

It would have been known immediately that the Russian explosion of August, 1953, was partly thermonuclear and that this test was many times as big as the Russians' previous explosions. If one assumes that following this Russian test the American program in the worst plausible world would have gone along just as it did in the actual world following President Truman's 1950 decision, then the U.S. would have set off the Mike explosion in April, 1956. A simple duplication of those earlier events at this later time, however, would have been unlikely. Any analysis of U.S. reactions to technological advances by the U.S.S.R. shows that the detection of the August 1953 event would have resulted in the initiation of a very large, high-priority American program to produce a bigger and better thermonuclear device. Such a program would undoubtedly have had broader support than the one actually mounted in the spring of 1950. Moreover, the general scientific and technological situation in which a hydrogen-bomb program would have been embedded in 1953 would have been significantly different from the actual one in 1950. For one thing, the kind of theoretical work in progress on the Super before President Truman's decision would have continued and would have provided a solider base from which to launch a crash program. In addition the booster program would presumably have continued along the path already set for it in 1948 (which included a test of the principle in 1951), and therefore in 1953 there would have been available some real experimental information concerning thermonuclear reactions on a smaller scale.

Last but not least, there had been great progress in computer technology between 1950 and 1953. When the real Mike test was being planned, fast electronic computers such as MANIAC and the first UNIVAC either were not quite operating or were in the early stages of their operating career. By a year or so later they were in full running order and much experience had been gained in their utilization, so that they would have been much more effective in connection with any hypothetical post-Joe 4 American crash program. For all these reasons it is plausible to assume that the U.S. would have arrived at something like the Teller-Ulam design for a multimegaton superbomb either in the same length of time or, even more likely, in a somewhat shorter period, say sometime between September, 1955, and April, 1956.

These dates bracket the actual date when the Russians arrived at roughly the same point in the actual world. A few months' difference either way at that stage of the program, however, would not have been meaningful. It takes quite a long time, typically several years, to go from the proof of a prototype to the deployment of a significantly large number of weapons based on it. Differences in production capacity would have played a much more important role than any small advantage in the date of the first experiment, and such differences as then existed surely favored the U.S. Hence even in the worst plausible alterative world the nuclear balance would not have been upset. Moreover, in the most probable alternative world the date the Russians would have arrived at that stage would have been delayed until well after the first large U.S. Mike-like explosion had showed them there was a better way; thus in this most probable case the U.S. would still have enjoyed a substantial lead.

In short, the common notion that has persisted since late 1949 that some sort of disaster would have resulted from following the Oppenheimer-Lilienthal advice is in retrospect almost surely wrong. Moreover, even if by some unlikely quirk of fate the Russians had achieved the Superbomb first, the large stock of fission bombs in the U.S. arsenal, together with the 500-kiloton all-fission bomb for those few cases where it would have been appropriate, would have adequately ensured the national security of the U.S.

This history and the conjectures about possible alternative pasts show that Oppenheimer, Conant, Fermi, Rabi and the others were right in their advice about the Super, and that they were right for the right reasons. They had correctly assessed the relative technological state of affairs, correctly judged the margin of safety inherent in the situation and correctly projected the ability of the U.S. to catch up rapidly if that should become necessary. The national security of the U.S. did not require the initiation of a high-priority program to develop the Super. It was therefore entirely appropriate to attempt to use the first Russian atomic explosion as a lever for reopening the entire question of nuclear-arms control.

The authors of the report could not, of course, predict the details of the alternative chronologies outlined above, and they did not try to do so, but they could and did correctly assess the general situation and the limits of the probable futures inherent in it. The large rate of production of fissionable material already in effect, the planned expansion in that rate, the resulting immense stock of fission weapons forecast for the early and middle 1950's and the existence of an entirely adequate means for delivering those weapons guaranteed that even the sudden surprise introduction of a few superbombs by the U.S.S.R. could not really upset the balance of power. The situation was reinforced by the projection, which proved to be correct in the King shot, that if need be the power of the World War II fission bombs could be multiplied up to the megaton range simply by more astutely employing the techniques and materials already known and available.

In the course of presenting its general admonition not to proceed with the crash development of the Super, the Oppenheimer committee made certain specific predictions about it. An examination of these predictions shows that they stood the test of time fairly well.

In their discussion of the superbomb the committee members said that "an imaginative and concerted attack on the problem has a better than even chance of producing the weapon within five years." Four years and four months later Bravo, the first practical American thermonuclear weapon, was tested at Bikini. Given the unknowns and uncertainties existing at the time, that is a remarkably accurate prediction. They went on to say that "once the problem of initiation has been solved, there is no upper limit...except that imposed by the requirements of delivery." That also seems to be the case. The largest bomb exploded so far (by the Russians in 1961) is said to have been some 58 megatons, four times the size of

Bravo, and there is every reason to believe bombs could indeed be made even larger than that.

The report also said that there "appears to be no experimental approach short of an actual test which will add to our conviction that a given model will or will not work" and that "many tests may be required before a workable model has been evolved." History has borne out the first part of the prediction. A quarter of a century had to pass and other inventions had to be made before thermonuclear explosions were produced on a laboratory scale by means of lasers, and even those are probably not closely relevant to the superbomb problem. The second part of the prediction turned out to be less precise. The number of U.S. tests needed to develop and check out a bomb was three: George, Mike and Bravo. The Russians needed only two tests, but they had an invaluable piece of information that was not available to the American workers: the sure knowledge that both small and large thermonuclear explosions were really possible. These numbers were very probably smaller than the "many" the Oppenheimer committee had in mind, but even so they were in each case sufficient to provide the other side with an adequate early warning that thermonuclear work was in progress.

Another interesting and perceptive technological prediction is contained in the report's statement about "very grave contamination problems which can easily be made more acute...by surrounding the deuterium with uranium." The very high levels of radioactive fallout associated with large hydrogen bombs do in fact result from such use of uranium. The very first test of a practical superbomb, Bravo, produced a blanket of fallout that evidently contributed to the death of one innocent bystander (the radioman of the *Fortunate Dragon*, a Japanese fishing ship) and came within a hair's breadth of killing hundreds of Marshall Islanders living on two nearby atolls. The fallout accident in turn provided the initial spark behind the movement to ban nuclear-weapons tests that ultimately led to the Partial Test-Ban Treaty of 1963.

The foregoing account is, I think, enough to show that the Oppenheimer committee's advice was sound, but it may not be enough to show unequivocally that President Truman should have taken this sound advice. The President, unlike the AEC commissioners and their advisers, had to take into account a broader array of information and political ideas than those discussed in detail here. The overall intensity of the cold war was increasing, Mao Tse-tung and Joseph Stalin had proclaimed the Sino-Soviet bloc and many important Republicans were withdrawing or modifying their support of the bipartisan foreign and military policies that had been in effect since the beginning of World War II. As the fall of 1949 wore on and the arguments about the Super began to leak out from behind the curtain of secrecy, those opinions favoring the Super were, in the overall context of the time, both simpler and more widely persuasive than those opposing it. There can be little doubt that Congressional and public opinion was beginning to come down heavily on the side of a strong response to the first Russian atomic-bomb test, and building the Super seemed to many to be just the kind of thing to keep the Russians in their place. President Truman, a professional politician, could therefore have concluded that rejecting the Super and running even a small risk of being second best was politically too difficult an alternative. Moreover, his decision to proceed with the Super, made on January 31, 1950, was based on the advice of the special committee of the National Security Council charged with studying the matter. Those committee members responsible for international relations (Secretary of State Dean Acheson) and national defense (Secretary of Defense Louis A. Johnson) strongly supported going ahead; the only reservations were expressed by the one committee member who was not responsible for those elements of national-security policy, namely Lilienthal, chairman of the AEC.

Nonetheless, it now seems clear to me in retrospect that President Truman should have taken the advice of the Oppenheimer committee; he should have held back on initiating the development of the Super while making another serious try to achieve international control over all nuclear arms, particularly the Super. The benefits that could have flowed from forestalling the Super altogether were incalculable; the chances of succeeding in doing so were small, but so were the risks in trying. It was certainly one of the few opportunities, and as Lilienthal said then, it may have been the last good opportunity to base American foreign policy on something better than reliance on weapons of mass destruction or, as it is now phrased, on the prospect of "mutual assured destruction."

Eisenhower's Other Warning

In his farewell address, President Eisenhower issued two warnings to the American people. The first of these is very well known, it fits easily into a variety of ideological frameworks, and it is often quoted or paraphrased. We must, he said, be wary of "the acquisition of unwarranted influence, whether sought or unsought, by the military-industrial complex."

Eisenhower's second warning is much less well known, it is not so easily understood, and it is seldom quoted except by specialists studying the Eisenhower administration. After noting that research played an increasingly crucial role in our society and that the ways in which it was conducted had changed radically in recent years, Eisenhower said, "Yet in holding scientific research and discovery in respect, as we should, we must also be alert to the equal and opposite danger that public policy could itself become the captive of a scientific-technological elite." Even on those occasions when some scholar does recall this second warning, it is common to ascribe both the words and the ideas to the President's speech writer. This is probably correct as far as the words are concerned, but I personally knew Eisenhower and his concerns well enough to be certain in my own mind that the ideas were his own.

To understand this second warning, it is necessary to recall its context. This context consisted of the events that took place during the forty months from the launching of Sputnik to the end of his administration. The particular segment of the "scientific and technological elite" that he had in mind consisted of the hard-sell technologists who tried to exploit Sputnik and the missile-gap psychosis it engendered. We should be wary, he said, of accepting their claims, believing their analyses, and buying their wares. They and their sycophants invented the term "missile gap," they embellished that simple phrase with ornate horror stories about imminent threats to our very existence as a nation, and they offered a thousand and one technical delights for remedying the situation. Most of their proposals were expensive, most were complicated and baroque, and most were loaded with more engineering virtuosity than good sense. Anyone who did not immediately agree with their assessments of the situation and who failed to recognize the necessity of proceeding forthwith on the development and production of their solutions was said to be out of touch with reality, technically backward, and trying to put the budget ahead of survival.

The claims of such people that they could solve the problem if only someone would unleash them carried a lot of weight with the public and with large segments of the Congress and the press. Other scientists and technologists had performed seeming miracles in the recent past, and it was not unnatural to suppose that they could do it again. It seemed that radar had saved Britain, that the A-bomb had ended the war, and that the H-bomb had come along in the nick of time to save us from the Russian A-bomb. On the home front, the relatively recent introduction of antibiotics had saved our children from the scourges of earlier times, and airplanes and electronics had become capable of carrying us, our words and our images great distances in short times. Scientists and technologists had acquired the reputation of being magicians who had access to a special source of information and wisdom out of reach of the rest of mankind. A large part of the public was therefore more than ready to accept the hard-sell technologist's view of the world and to urge that the government support him in the manner to which he wanted to become accustomed. It seemed as if the pursuit of expensive and complicated technology as an end in itself might very well become an accepted part of America's way of life.

But it was not only the general public that believed the technologists understood something the rest of the world could not. Many of the scientists and technologists themselves believed that only they understood the problem. As a consequence, many of them believed it was their patriotic duty to save the rest of us whether we wanted them to or not. They made their own analyses of what the Soviets had done. They used their own narrow way of viewing things to figure out what the Russians ought to do next. They then argued that since the Russians were rational (about these things anyway), what they ought to do next is what they must in fact now be doing, and they then determined to save us from the consequences of this next real or imaginary Russian technological threat. The Eisenhower Administration was able to deal successfully and sensibly with most of the resulting rush of wild ideas, phony intelligence, and hard sell. But some of these ideas did get through, at least for a while. Beyond that, dealing with self-righteous extremists who have all the answers—and there were many among the scientists and technologists at the time—is always annoying and irritating.

As we now know, the commonly baroque and occasionally bizarre technological ideas urged on us in those years were in fact a portent of things to come. Weapons systems and other high technology devices have become still more complex in the years since Eisenhower's farewell address. And this complexity is creating serious social and political problems of the general kind that Eisenhower warned us about. Today, there are even more people who tell us that because major weapons systems are so complicated only weapons experts can decide if they are needed, that only those in on all the secrets and up on the most arcane elements of operations analysis can tell us whether arms control and disarmament is good or bad, and that only nuclear experts are fit to decide whether, when, and where nuclear power plants should be built. There are today many scientists and engineers, and many members of the general public as well, who believe that basic issues like these are simply beyond the ken of the people and their elected representatives, and that public policy concerning such matters should indeed be made by a scientific technological elite. Eisenhower's second warning is even more pertinent today than it was when he made it.

Let me now turn to a more personal matter. Eisenhower's Second Warning profoundly affected my own life in two ways, one substantive and the other fortuitous.

As fate would have it I worked fairly closely with Dwight D. Eisenhower during the last three years of his Presidency, first as a member of the Science Advisory Committee he set up immediately after Sputnik under the chairmanship of James R. Killian, Jr., and second as the first Director of Defense Research and Engineering, a new position created in 1958 as another part of the response to Sputnik. In these positions, I was directly concerned with precisely those scientific and technological programs in which the President himself was most involved and my own view of the world gradually changed as I came to see and understand the overall situation in which we found ourselves. I had gone to Washington a technological optimist, full of confidence in the technological fix. I came away three and a half years later gravely concerned about the all too common practice of seeking and using technological palliatives to cover over serious persistent underlying political and social problems. In particular, I became convinced of the futility of always devoting our main efforts to finding a technical solution to the problem posed by the steady decrease in our national security that was being brought about by the spread of high technology weapons throughout the world. This, it seemed to me, was not only futile but basically absurd, because nearly all of the weapons which in the hands of others were (and are) threatening our national security, and indeed our very existence, had been invented or perfected by us in the first place. In sum, my views on the relationship between technology and security did not arise out of Eisenhower's warnings, rather his warnings and my views both arose out of the same set of circumstances, but his formal warnings did very much help to crystallize my views on the subject. I found it very reassuring that the Commander-in-Chief, a professional military man himself, shared my own growing doubts about the value and efficacy of placing such a relatively high priority on finding technical solutions to what were really political problems.

In addition to this substantive side of it, it also happened that after leaving the Presidency, Eisenhower spent his winters at a California desert resort less than one hundred miles from my home, and I called on him there on several occasions to pay my respects. Much of our conversation on those visits was devoted to the two warnings. He told me quite specifically that he had just two purposes in mind in making his farewell address. One was to say goodbye to the American people, whom he deeply loved. The other was to bring before the people precisely these warnings. The rest of the speech, he said, perhaps with some exaggeration, was there simply to make the whole thing the appropriate length for a farewell address. I asked him to explain more fully what he meant by the warnings, but he declined to do so, saying he didn't mean anything more detailed than what he had said at the time. I knew him well enough to understand what he meant: these warnings were not the result of a careful, methodical analysis; rather, they were the product of a remarkable intuition whose power has generally been underestimated. I pressed this line of questions further by asking him whether he had any particular people in mind when he warned us about "the danger that public policy could itself become the captive of a scientific-technological elite." He answered without hesitation: "(Wernher) von Braun and (Edward) Teller." A foreshadow of Star Wars, more than twenty years ahead of time!! At one point, in response to my persistent questions, he invited me to draft a couple of pages expanding on this idea that he might incorporate in his next book.* I was flattered, but I didn't do so, mainly because I had never done any expository writing of that sort before. However, I did not forget his invitation and a few years later when I wrote *Race to Oblivion*, our conversations and his suggestion that I write something about the problem of the technological arms race was very much in my mind. In fact, the pro-

---

*Eisenhower, D.D., *Waging Peace, 1956-1961*, Garden City, N.Y., 1965.

logue to that book consists of a discussion of the warnings in his farewell address, much like the one presented here.

Eisenhower's warnings which were based largely on his intuition, pointed up very real and extremely serious problems. If we forget or downgrade his warnings, it will be to our peril. Most of what I have written since that time has been designed to help keep these warnings before us, and to contribute to finding some way of coping with the issues raised by them.

NEGOTIATING
LIMITS

Negotiating and the U.S. Bureaucracy

U.S.–Soviet negotiations are very complex because every negotiation actually involves several negotiations. I will focus here in particular on the bureaucratic negotiating process within the U.S. government, and how it influences negotiations with the Soviets. One simply cannot understand what the United States does, or even more importantly what the United States does not do, without understanding the bureaucratic processes that underlie those events and non-events.

In a bilateral negotiation, there are really three distinct negotiations. There is a negotiation among all the interested parties in Washington. There is probably something like that in Moscow, although it is a bit mysterious. And then, of course, there is a negotiation between the delegations of the two countries in Geneva.

In the case of the Comprehensive Test Ban (CTB) negotiations, the situation was further complicated by the inclusion of the British. It was a trilateral negotiation. This added two more negotiations: a negotiation among the British in London, and a secondary negotiation between the United States and Britain in which we sought, not always successfully, to work out a common position to present to the Russians. The Russians always referred to the United States and Britain as one side, which of course we were. But that did not mean that our positions were always identical, and therein lie some of the problems with multilateral negotiations.

The Role of the Backstopping Team in the Comprehensive Test Ban Negotiations

The backstopping team is comprised of representatives of several agencies, each of which wishes to have some input into the process, and each of which has some expertise that might be needed or some viewpoint that should be heeded.* For CTB, the backstopping team included the usual agencies: the Department of Defense (DOD), the CIA, the Joint Chiefs of Staff (JCS), the Department of State, and the Arms Control and Disarmament Agency (ACDA). It also included the Department of Energy (DOE) because DOE, not DOD, is responsible for conducting nuclear weapons tests. Thus, the CTB negotiation, whose purpose was to stop nuclear testing, was supposed to stop a major activity of the Department of Energy.

In the CTB, and to some extent in the antisatellite (ASAT) negotiations, the situation was different from many other negotiations in one important respect. In the CTB case, there were certain agencies that were not interested in simply contesting or modifying certain details of a particular proposed test ban; they wanted no test ban at all. They flatly opposed the president's objectives. These agencies often couched their objections in terms of modifications to a specific proposal, but in reality they flatly opposed a test ban. The DOE opposed the test ban because it was their activity that was to be banned. The Joint Chiefs opposed a test ban because they genuinely felt, and were never convinced otherwise, that testing was essential to maintain the reliability of our nuclear stockpile. Thus, all other issues, such as verification, were really secondary. The agencies did not want a comprehensive test ban at all, and no simple modification could satisfy them.

The case of the ASAT negotiations was similar, but less extreme. Within the DOD, the Air Force was interested in pursuing antisatellite negotiations. They had certain reservations about the idea, but nevertheless expressed interest. On the other hand, the Navy, or at least elements of it that entered the debate, was flatly opposed to a ban on antisatellite weapons. Regardless of what the Russians do, they argued, the United States

---

* There was considerable discussion of backstopping. One person asserted that the U.S. government needed to be better organized for negotiations and that National Security Council control of backstopping committees could assure this. A controversial suggestion to resolving interagency squabbling over negotiating positions was the appointment of a special high-level official to work full time to coordinate among the components of the backstopping effort. One speaker responded that one person doing this for each negotiation was too much and perhaps the director of the Arms Control and Disarmament Agency could fulfill the coordinating role. Another participant said we needed more non-political expertise on the Soviet Union, which could be especially helpful in crises. Someone responded that this would be hard to develop because diplomats and others did not like long stays in Moscow and anyway it would not guarantee a more rational view of the Soviet Union.

must have antisatellite weapons in order to cope with Soviet sea surveillance systems. By a curious quirk, the representative of the Office of the Secretary of Defense (OSD) on the backstopping committee was an admiral who adhered to the Navy position. He worked for OSD, not the Navy, but he was the OSD representative on the backstopping committee and was personally opposed to a ban on ASATs.

As noted above, the backstopping committee for the CTB included representatives of the DOE and the JCS, agencies that flatly opposed any positive result. Furthermore, the individuals were in some cases even more strongly opposed than their agencies. That was significant because the process of formulating negotiating instructions is essentially bureaucratic. And in any bureaucratic process, the person who wants only to obstruct the process has an advantage. There is nothing one can do to satisfy him. He does not want some particular change. He simply want no agreement at all. The natural advantage of a bureaucrat in that position is easy to exploit. One does not even have to be a particularly smart bureaucrat to exploit it, although a smart bureaucrat can do so more effectively.

There are two common tactics for exploiting this advantage. One is to propose that the problem of the moment is much more complicated than it appears and that it requires further study. The other is to insist that the problem is beyond our level of competence and must be resolved at a higher level. In the case of the CTB backstopping group, a higher level meant the cabinet-level committee, called the Special Coordination Committee during the Carter administration and the Committee of Principles during the Eisenhower administration. Even in good times, any question referred to the cabinet level would take weeks or months to resolve. And after the hostage crisis, Afghanistan, and the presidential primaries, the Special Coordination Committee never again resolved anything in regard to the CTB.

The following specific examples illustrate use of the first tactic: saying that a problem is too complicated and requires further study. The first example also illustrates the problems of three-party negotiations. About halfway through the CTB negotiation, each side had advanced major proposals concerning the placement of ten special seismic stations for monitoring compliance with a test ban on the territories of the other party. Then, all delegations returned to their capitals to work out details of the locations of these stations in the United States, Britain, and the Soviet Union.

The individual proposals were then jointly discussed at a cabinet meeting in London that happened to convene during a rare London snowstorm. The dustmen, who put sand on the roads to facilitate driving, were on strike. As a result, the meeting was late, it was short, and the main topic was how to clear the roads. There was little time to discuss a comprehensive test ban, but it was decided that the United Kingdom, because it is a small country, should have only one seismic station, unlike the United States and the Soviet Union, which would have ten each. This decision came as a surprise to everyone, and brought the negotiation to a standstill.

To move the process off dead center, I immediately proposed that Secretary of State Cyrus Vance meet with British Foreign Minister David Owen to work out a compromise somewhere between the British proposal of one station and the Soviet proposal of ten. Secretary Vance was just beginning preparations for that meeting when the delaying tactic was employed. The representative of the Joint Chiefs said that Vance should not independently discuss the question of seismic stations with Owen because it had important military implications that required further study. The tactic worked. The resulting study delayed a Vance–Owen meeting for several months.

The study finally concluded that in fact Vance should meet with Owen on this issue. But by the time that conclusion was reached, there had been an election in Britain. Prime Minister James Callaghan had been replaced by Margaret Thatcher, who considered the CTB a bad idea. She was willing to follow the American lead in pursuing a CTB. However, she felt that if one seismic station was enough for Callaghan, it was certainly enough for her. And so we were stuck. The two-month delay was totally effective; it removed the possibility of working out a more acceptable position.

Ironically, ACDA, which supported a test ban, also helped to slow things down in this case. A few purists in ACDA argued that we should not allow the British to settle for an intermediate position. They insisted that the British must accept ten seismic stations. If the British want to play in this league, they argued, they must play by the rules. Thus, the ACDA purists also delayed the process by adhering to an extreme view on the other side.

A second example of bureaucratic-induced delay concerns one of the more bizarre situations I faced. In late 1978, we invited the Soviets to send a delegation to the United States to look at our proposed seismic stations, just to see what they were like. We hoped to allay Soviet fears by removing some of the mystery. As is usual with the Soviets, they took a long time to answer. However, more than six months after receiving our original invitation, they finally did agree to come. Eager to get the program ready, I immediately reported the Soviet acceptance to Washington.

Unfortunately, the DOE raised two issues. First, they asked who was going to pay for the visit. It was a matter of $25,000, essentially for hospitality. The norm in such exchanges is that the Soviets pay for us when we visit there, and we pay for them when they visit here. This $25,000 issue blocked progress for almost two months because the money did not appear anywhere in the budget. DOE representatives claimed that there could be trouble if the House Armed Services Committee learned that we were spending unbudgeted funds. The people who raised the issue were delighted that there were problems, but they adopted a pose of helpfulness. Nevertheless, they insisted that the issue be resolved before reconfirming the invitation and completing the plans for the visit.

The second issue DOE raised was that scientific exchanges with the Soviet Union must be based on agreements either between the two science academies or between various agencies in the government. In this case, they argued, there is no agreement to exchange data on seismology, and we must conclude such an agreement before the Soviets can visit our stations. This was an awkward situation indeed. We had already invited them. Six months later they had agreed to come. I had told them that we were eagerly awaiting their visit. Then I received instructions from Washington not to arrange anything. So I spent the next two months telling the Russians that there were some minor administrative questions to be settled. We finally did resolve these issues. It delayed their visit for over two months, but the Soviets did come.

A third example of delay concerns a proposal to put a seismic station in Obninsk, a city near Moscow that has a Soviet seismic observatory. Most U.S. officials thought it would be quite worthwhile to have an operating American station in the Soviet Union. It is difficult to find anything wrong with the idea. We could acquire a lot of information from such a station. Also, the Soviets would learn more about our equipment and possibly overcome some of their fears about placing American equipment on Soviet territory. In addition, we would get some seismic data from a site that was already thoroughly understood.

But when the full backstopping group considered the proposal, the group's negative members argued that we should not even propose putting a station in Obninsk without first developing a complete plan for its use. They raised the following questions: How will we get it there? Who will install it? What kind of data will we get back? What will we do with that data? We must study the problem more thoroughly, they argued. That tactic delayed our presentation of a proposal to the Soviets for more than three months.

We advanced the proposal on December 5, 1979, during the period between the capture of the U.S. Embassy in Tehran and the Soviet invasion of Afghanistan—a singularly difficult time. The proposal came just at the end of the round. The Soviets indicated that they were unsure about the proposal, but they would grant it "the consideration it merited," and would discuss it back in Moscow. Our next meeting was after the invasion of Afghanistan. We were instructed to proceed very slowly. Some of us understood that as an order to move faster because, by that time, we were not in fact moving at all. But we proceeded even more slowly.

As one reaction to the Afghanistan invasion, the president further restricted technology transfer. The seismic station is a technological device, a seismometer that, with the sole exception of one cryptologic device involving some modern technology, uses straightforward, simple technology. Moreover, the cryptologic device could have been omitted at the Obninsk site. There was no real technology transfer problem, but the same people tried to anchor their objections on the newly imposed policy. Now they argued that we thoroughly study every component of the device to determine whether it should be placed in the Soviet Union.

In sum, we had a great idea: to put a seismometer in the Soviet Union as part of an experiment before the final CTB treaty would be signed. It would have moved the process forward, and we would have obtained some data. But the idea was blocked by those who argued we should study the technology transfer implications of the issue. The bureaucratic delay was then transformed into a permanent roadblock by events in Afghanistan, Tehran, and the presidential primaries. The notion that a cabinet-level committee was needed to settle the issue brought the process to a complete stop, because it was not possible to convene a cabinet-level meeting on such a trivial issue under the prevailing circumstances.

I returned to the subsequent round of negotiations with a curious set of instructions. I was told not to mention the December 5 proposal unless the Soviets mentioned it first. Evidently, they had exactly the same instructions. We had struck an impasse. Thus, we ended the negotiations one year later without ever again mentioning our proposal, whose purpose was, in my judgment, entirely beneficial to the United States.*

The Evolution of U.S. CTB Positions

Strong opposition within the executive branch led to another peculiarity of the CTB negotiations. The path that the negotiations followed was entirely

---

* One participant suggested that the "nay-sayers' in the bureaucracy were strengthened in the latter part of the 1970s by a conservative shift in U.S. domestic politics. He said we needed better political timing in our negotiations to know how far we can go in our arms control positions. Others added that we needed strategies for accomplishing our arms control objectives while they are still relevant, noting the obsolescence of SALT II by the time it was finally signed.

contrary to folklore about how arms control negotiations usually proceed. Specifically, the American position steadily hardened throughout the negotiation, while the Soviets made occasional concessions that lagged behind changes in the American position. This situation is contrary to the usual view that the Soviets stand firm while we gradually make concessions to them.

We began with the position that a CTB should be forever. It should, of course, allow for withdrawal, and it should allow for review, as do other arms control agreements. But our position was that the purpose was to permanently ban nuclear testing, assuming the necessary conditions were met. The Soviets, however, wanted only a temporary ban. They argued that the Chinese and French would eventually have to participate for the test ban to be a viable enterprise, and without Chinese and French participation, the Soviets were unwilling to permanently ban nuclear tests. The United States did not want a treaty that explicitly depended on what the French and Chinese might do. We wanted a permanent treaty that provided both sides the option to withdraw.

Early in the Carter administration, it became evident that the opposition to a test ban was stronger than the president had anticipated. So he began indirect negotiations with those senators who were predisposed against ratification of a CTB. In particular, Carter sought the support of officers and officials who had influence with those senators. He also began to make compromises designed to keep the support of those who were wavering and to obtain the support of others who were opposed. In his first such concession, Carter agreed to scrap the idea of a permanent test ban in favor of a five-year ban.

Unfortunately, Carter ordered this switch in the U.S. position just after the Soviets had come around to agree to our initial proposal for a permanent test ban. But the Soviets again followed our lead and agreed to a five-year ban, whereupon the president made another concession. Under pressure from the DOE and the Defense Nuclear Agency, he decided that a three-year ban was the longest we could tolerate. The Soviets never agreed to three years, but they did agree to continue discussing other issues, while postponing the question of the duration of a test ban.

Another fundamental U.S. position going into the negotiations was that a nuclear test ban was a very simple thing. There were to be "no nuclear weapons tests," and no nuclear weapons tests meant none. After about one year, and under pressure from advocates of continued testing, the administration decided that it should formulate a more precise definition of a "nuclear test." It agreed that there are certain experiments that really are not nuclear tests, but that might look like nuclear tests, and that might be necessary and legitimate to conduct. These were called "permitted experiments." The U.S. negotiators formally introduced this idea of permitted experiments about one year into the negotiations, promising to present definitions of such activities shortly. In virtually every round, the Soviets would ask for our position on permitted experiments. But three years later, when the negotiations finally terminated in 1980, we still had not developed a definition. That particular point went unsettled.

The initial U.S. position on the potential manufacturers of the seismic stations also hardened as the negotiations continued. At first, the United States said that it did not matter who manufactured the stations. At one time, Ambassador Paul Warnke said they could be manufactured anywhere, perhaps in Japan, even in Mongolia. What matters, he argued, is that the stations meet certain technical performance criteria, and we should negotiate those criteria with great care. So we proceeded to negotiate the technical parameters describing these seismographs. We said that we knew how to make them, and we wanted the Soviets to see our equipment, but we did not care where the seismographs were made.

Then, after two years of negotiations, in another attempt to compromise with those opposed to a CTB, the president decided that the seismic stations to be placed in the Soviet Union must be manufactured in the United States. The Soviets never rejected this new proposal outright, but they never accepted it either. And in one rather heated exchange, the Soviet representative accused me of treating the Soviets as if they were a sixth-rate Arab nation, a puzzling but vivid analogy.

We changed our position on other issues too. At first, the U.S. said that it must receive data from seismic stations in the Soviet Union "in a timely fashion." Halfway through the negotiation, we decided that we must receive the date "in real time."

Also, we began with the notion that the preamble would present no problems. We have written numerous preambles in the past and, by referring to previous texts, we assumed it should be easy to elaborate a preamble for this treaty. Again, two years into the negotiation, the American position changed. The bureaucratic opponents of the CTB said that we had been promising too much in past preambles. We had been making "pie in the sky" promises we could not keep. Thus, it was concluded that the preamble should be negotiated with great care to avoid making excessive promises. Despite both British and Soviet proposals of various alternate preambles, we refused even to discuss the preamble after the first six months. Only during the last six months, when there was nothing else to do, did we resume preamble discussions.

With regard to the natural advantage of the "nay-sayers," there are only two possible solutions. First, if the president is interested enough to become personally and continuously involved, he can simply regularly overrule anyone who is trying to block any agreement at all. The other solution is to deny them coequal status after negotiations have begun. It is necessary, for political and other reasons, that all interested agencies be involved in a coequal basis in deciding whether there ought to be a negotiation. But once that is decided, it is self-defeating to grant coequal status to agencies and people whose sole real goal is to stop the negotiation or cause it to fail.

Other Observations

I will conclude with several observations on negotiations in general. One has to do with the role of visitors. While I was negotiating the CTB, we had, among others, two Washington visitors—one from ACDA and one from the National Security Council staff. Each of them came to familiarize themselves with the process. Beyond that, they had no specific purpose. However, the Soviets were convinced that these visitors were special emissaries who came to float trial solutions to a particular issue on which we were deadlocked, and they so interpreted remarks that were made during informal, essentially social, visits.

In one particular case, the Soviets interpreted the visit as a signal that we were preparing to compromise on the issue of how many seismic stations to put in the United Kingdom. As a result, they made a minor, tactical concession. They agreed to talk about something that they had previously refused to discuss. So we discussed the issue and thanked them for their concession. They latter said that they felt we had deliberately tricked them into making this concession by parading visitors to create the impression that we were ready to compromise.

That raises the issue of the Soviet approach to compromise. One could argue that the Soviets use rigidity and stubbornness as a tactic to elicit U.S. compromises. Be that as it may, their willingness to make concessions does not hinge only on our firmness. It also depends on their assessment of our seriousness about the negotiating process. In the Soviet viewpoint, the worst mistake that one of their negotiators can make is to offer a concession that is not reciprocated. Thus, if the Soviets decide that we are not serious, they will refuse to make real concessions on even the most trivial matters.

Midway through the test ban negotiation, the United States became unwilling to discuss issues that we previously wanted to discuss such as the preamble and permitted experiments. We steadfastly and continually refused to discuss those issues. Meanwhile, Leslie Gelb, shortly after he left the State Department, published an article in Foreign Policy, corroborated by other stories in the American press, saying that the CTB was going nowhere.* At that point, the Soviets became convinced there was no chance for a test ban negotiation to succeed, and from then on, they refused to make even the slightest concession.

---

* Leslie H. Gelb, "A Glass Half Full...," Foreign Policy 36 (Fall 1979): 21-32.

Comprehensive Test-Ban Negotiations

Before Carter

Since the bombing of Hiroshima all American presidents had actively sought means to contain, stop, and reverse the nuclear arms race. Jimmy Carter differed from his predecessors not in kind but only in degree. He tried harder than they and explored the broadest set of possibilities. In one of modern history's all too common ironies, he accomplished the least. Events over which he had little control ultimately prevented him from adding very much to the limitations already worked out and put into place by the efforts of Eisenhower, Kennedy, Johnson, Nixon, and Ford.

The idea of a nuclear test ban as a first step in containing the arms race emerged in the mid-1950s. The first concrete accomplishment was the nuclear test moratorium of 1958–61, achieved under the leadership of Eisenhower and Khrushchev. The purpose of the moratorium was to provide a political climate suitable for working out a formal treaty permanently banning such tests. Tripartite negotiation to this end, including the United Kingdom as well as the United States and the Soviet Union, were in almost continuous session from october 31, 1958, to January 19, 1962. Internal disputes among the Americans studying the issues in Washington, combined with some hard-nosed attitudes on the part of the Soviets in Geneva, delayed the negotiation of a treaty. Finally, in 1961, the sudden resumption of nuclear testing by the Soviets brought the whole process to a total, but temporary, halt.

The dangers demonstrated by the 1962 Cuban missile crisis inspired Kennedy, Khrushchev, and Prime Minister Harold Macmillan of Britain to make another try. As before, the single most difficult issue was the monitoring of underground tests. Small nuclear tests made seismic signals much like those generated by earthquakes, and the latter were very common in the USSR American specialists could not agree even among themselves, much less with the Soviets, on the means and prospects for coping with this problem.

Tripartite negotiations in Moscow in 1963, led on our side by Ambassador Averell Harriman, achieved a treaty banning all tests except those underground. In addition to banning tests "in the atmosphere" in outer space, and under water," the treaty pledged the parties to continue to seek "to achieve the discontinuance of all test explosions for all time

and "to continue negotiations to that end." In order to be sure of getting the support needed for ratification by the Senate, President Kennedy agreed to support a vigorous program of nuclear weapons development, including whatever underground tests were necessary to that end.

During the Johnson administration two additional major multilateral arms control treaties were elaborated. One, the Outer Space Treaty, banned the deployment of "nuclear weapons or any other kind of weapons of mass destruction" in space or on "celestial bodies." The other, the Treaty on the Nonproliferation of Nuclear Weapons, was designed to prevent the spread of nuclear weapons to states that did not already possess them. Its own preamble recalled the pledge in the preamble of the Limited Test Ban Treaty calling for a comprehensive ban. More important, Article 6 of the Nonproliferation Treaty called on the parties to pursue negotiations in good faith on effective measures relating to the cessation of the nuclear arms race at an early date and to nuclear disarmament."

Richard Nixon, in addition to the more important SALT I Treaty, also directed the negotiation of a treaty setting the upper limit on the permissible size of underground explosions at 150 kilotons, more than twelve times the power of the Hiroshima bomb. Called the Threshold Test Ban Treaty (TTBT), it was signed by Brezhnev and Nixon a month before the latter's resignation.

Gerald Ford placed highest priority on continuing but he also managed to add one more treaty to those tests. Known as the Peaceful Nuclear Explosions Treaty (PNET), it was signed by Brezhnev and Ford in 1976. It is, in effect, a supplement to the TTBT, designed to eliminate the possibility that certain allowed "peaceful nuclear explosions" might be used as a cover for weapons tests exceeding the 150-kiloton limit. Like his predecessors, Ford in principle supported the goal of a comprehensive test ban treaty, but he did not conduct negotiations to that end. As he later put it, "You can only handle so many things on your plate at one time, and a test ban was not our highest arms control priority."[1] Like Johnson and Nixon before him and Carter after him, Ford put the highest priority on limiting, and ultimately reducing, the number and types of strategic weapons—that is, on the SALT process. Ford's summit meeting with Brezhnev in 1974 was devoted to this process, and so were the talks that took place under his direction at Geneva in 1975 and 1976.

First White House Review

Arms control is, as Paul Warnke once noted, an unnatural act. A new president must therefore always review the situation very carefully before proceeding. This was especially true for Carter, who was widely known to be eager to move in this area. Such a review provides the new president and those of his key associates who are inexperienced in the matter with the knowledge necessary for making sensible decisions. It also serves a more political purpose by helping to assure those who have substantial doubts about arms control that their concerns will at least be heard.

As usual, the review, or "interagency study," involved people from several different agencies and levels of government. In addition to the White House staff, these included the Departments of State and Defense, the Energy Research and Development Agency (which at that time funded the work of the nuclear laboratories), the ACDA, the CIA, the Joint Chiefs of Staff, and the Defense Nuclear Agency. The group's initial report, known as PRM-16 (Presidential Review Memorandum No. 16), confirmed Carter's intuition and led him to take further steps necessary to set in motion the trilateral Comprehensive Test Ban negotiations of 1977–80.

The interagency study did not end the review process. Further studies to explore and elucidate the problems inherent in the negotiations themselves continued in and among the same agencies. It was my good fortune to be involved in two of them, one in the White House and one in the Pentagon.

The principal White House study was directed by Carter's science adviser, Frank Press, a geophysicist from MIT. Frank assembled a panel of experts including Hans Bethe, Richard Garwin, Carson Mark, W.K.H. Panofsky, Jack Ruina, and me. In addition, the two nuclear laboratory directors, Harold Agnew of Los Alamos and Roger Batzel of Livermore, sat regularly with the panel.

By that time, spring 1977, the argument over the utility of a test ban had come down to making a judgment about the relative value of two quite different factors, one of which weighed in on each side.

The main argument in favor of a test ban was that it was a necessary element of our nonproliferation policy. The Nonproliferation Treaty of 1970 called on the nuclear weapons powers to negotiate in good faith to end the nuclear arms race, and that was widely interpreted to require a comprehensive test ban as an early step. In 1975 the first quinquennial review of the nonproliferation regime focused major attention on this point. To be sure, nuclear proliferation had proceeded much more slowly than had originally been expected, but this situation could change quickly. A change was especially likely if the superpowers continued to engage in "vertical proliferation," a phrase meaning the development and deployment of ever more varieties of nuclear weapons in their own arsenals.

The main argument against a test ban revolved around the issue of stockpile reliability. Nuclear weapons experts pointed out that these devices were built of both chemically active and radioactive materials, which steadily undergo changes that can adversely effect their performance. Occasional full-scale nuclear tests would be necessary in order to assure that old weapons still worked. Even rebuilt weapons would inevitably include small, supposedly harmless changes in their manufacture, and these, too, would have to be subjected to full-scale tests to assure performance. Opponents of a test ban also argued that we needed to continue testing in order to build safer and more secure bombs, to develop bombs properly optimized for new delivery systems, and to learn more about weapons effects. Perhaps more important, continued testing was said to be needed in order to preserve a cadre of weapons design experts at the laboratories. In the absence of an experimental program, the experts would gradually move into more dynamic areas; when and if we ever needed them again, it would take a long time to reassemble or regenerate them. The Soviet laboratories, it was argued, would not suffer the same kind of loss in a test ban environment.

We studied the problem of stockpile reliability thoroughly and, except for the lab directors, decided that the nuclear establishment's worries were exaggerated. In brief, we concluded that regular inspections and nonnuclear tests of stockpiled bombs would uncover most such problems and provide solutions to them. Moreover, the laboratories could, if they tried, find ways around those that might remain. Agnew and Batzel disagreed. The in-house staffs in the Department of Energy and the Defense Nuclear Agency concurred with the laboratory directors, and the higher authorities in those agencies accepted their advice in the matter. The Joint Chiefs of Staff, whose nuclear arm is the Defense Nuclear Agency, also accepted the conclusions of the working-level experts immediately responsible for such matters. They really had no other choice.

Secretary of Energy James Schlesinger also felt that Carter was making a serious error in pushing ahead with a comprehensive test ban. Schlesinger had previously been chairman of the AEC, director of the CIA, and secretary of defense, and his views were based on his experiences in those posts. In an attempt to dissuade Carter, he arranged to have the president meet with Agnew and Batzel so that they could explain to him why further tests were needed.

This direct intervention by the laboratory directors at the highest level eventually caused quite a stir in the University of California, a stir that persisted for many years. The regents of the University managed the laboratories, and the directors were responsible to them through the university president. Most faculty members favored a test ban. In addition, a long series of faculty consultations had shown that roughly half of the faculty felt the university should not be operating the labs. It was therefore no surprise that many faculty reacted negatively when they learned about what they regarded as unwarranted political intervention by persons who were, ostensibly, spokesmen for the university.

I believe that the lab directors acted responsibly and properly. The president of the United States can consult with whomever he wishes. The consultants, in turn, are morally bound to tell him the truth as they see it. And that is exactly the way it happened. I, and many other members of the faculty, disagreed with the lab directors both on the facts and on their policy implications, but that was beside the point. I knew them both very well, and I had no doubt they told the president the truth as they saw it. I cannot fault them for having done so.

Harold Agnew has reported that the president took their arguments seriously. Others in a position to know say that Carter was not, in fact persuaded by the substance of their arguments. He did, however, come to understand that the opposition to a test ban was deeper and stronger than he had realized. Pressing hard for a test ban could interfere with related objectives, particularly with SALT II. Carter therefore adjusted his priorities so that a comprehensive test ban would not come up for Senate ratification before SALT II could. He did not otherwise modify his drive for a comprehensive test ban. Everything I know confirms this second view, and not Agnew's, of this matter.

The other technical issue that absorbed our attention involved the threshold for detecting and identifying nuclear explosions in the Soviet Union. The same majority considered the threshold was low enough that we could detect any test explosion that could be of real use to the Soviets. The Department of Energy and laboratory representatives again disagreed.

In sum, the Press panel reconfirmed Carter's intuitive view that a comprehensive test ban was in our national interest and could be adequately monitored, and it reinforced his determination to proceed with negotiations.

An initial bilateral U.S.-Soviet meeting on the subject was held in Washington in July 1977. In October, trilateral negotiations among the British, the Soviets, and the Americans were initiated in Geneva. Paul Warnke, also simultaneously the director of the ACDA and the chief U.S. representative at most other arms control negotiations, headed the American team.

First OSD Review

Shortly after the Press panel started its review, I also began to work closely but informally with several individuals and groups charged with reviewing these same issues in the Pentagon. Among them were David McGiffert, assistant secretary of defense for international security affairs, and two of his aides, Walter Slocombe and Lynn Davis, all of whom were newly nominated to their posts by Harold Brown. Another was James Wade, a longtime high-level member of the Defense Department planning bureaucracy and a specialist in nuclear affairs. The new appointees all supported the President's quest for a CTD. Wase opposed it, and so did everyone else—civilian as well as military—on the nuclear side of the Pentagon.

I spent a lot of time working with Wade. Despite our differences over fundamentals, we were able to make joint studies of certain details of a test ban. Among other things we examined possible alternatives that might accomplish the same objective but be generally more acceptable in the Pentagon. These included threshold test bans limiting explosions to very low yields, in the kiloton range and even below. We also looked at various quota systems that might limit the number of tests per year to some very low number, perhaps just one. Such a regime might provide enough tests to handle the stockpile reliability problem while slowing the rate down so much that a serious nuclear arms race would no longer be possible. To put it differently, a very low limit would in effect put an end to so-called vertical proliferation.

None of the alternatives I could think of were of much interest to either the proponents or the opponents of a CTB. Those who believed testing was essential regarded them as totally inadequate for their requirements. Those who believed a test ban was the keystone for further progress in arms control felt the same. There were, it seemed, no in-between solutions. It had to be either a total comprehensive test ban or nothing. In this situation Harold Brown and those he had appointed to high positions in the Pentagon supported the president's objectives. Most of the military, including the Joint Chiefs of Staff and, essentially, the entire permanent civilian nuclear staff, opposed it openly and strongly.

I often discussed these issues privately with Harold, usually on Saturday afternoons, just after I had met with Wade. At one such meeting he asked me for my private view of a letter that had recently been addressed to the president by Dick Garwin, Carson Mark, and Norris Bradbury and made public by Senator Edward Kennedy.

That letter discussed the matter of stockpile reliability. In it the authors wrote that the key question to be answered was the following:

Can the continued operability of our stockpile of nuclear weapons be assured without future nuclear testing? That is, without attempting or allowing *improvement* in performance, reductions in maintenance cost and the like, are there non-nuclear inspection and correction programs which will prevent the degradation of the reliability of stockpiled weapons?[2]

Garwin, Bradbury, and Mark answered their question with an unequivocal yes.

The three authors were well qualified to discuss the matter. Bradbury had been the director of Los Alamos for more than twenty-five years following World War II. Mark had been head of the laboratory's Theoretical Division for the same period of time, and Garwin was Garwin—an outstanding expert on many aspects of military technology, including nuclear weapons. In my commentary on their letter, I first wrote that I agreed with the technical judgment presented there, but I immediately added that that did not settle the issue. What ultimately mattered was what the top leadership of our national security apparatus concluded and not what a small minority of former scientific leaders (myself included) might think. And my review of the situation made it clear that the "leadership does not have confidence in a stockpile in a hypothetical future in which testing has been banned for many years, and it is also clear that no amount of persuasion from the outside world would convert it from its current position." Even if a CTB could somehow be put in place, I felt, it would not be politically stable:

In my experience, the U.S. national security establishment, both the part currently in government and that just outside, is too high strung and nervous to live contentedly with a CTB. Horror stories would thrive in a CTB environment and agitation within the executive branch and in the general political arena would be widespread and constant....

I continue to believe that a CTB is in the best interests of the United States and the world, but I have doubts about the advisability of achieving a comprehensive test ban in a single step.... I am inclined to favor an intermediate step consisting of a simultaneous limit on yield (say 10KT) and on testing rate (say one or two per year) coupled with forthright presidential statement to the effect that we have stopped developing new nuclear weapons but we must from time to time conduct proof tests in order to assure the condition of our current stockpile....

A Call from George Seignious

In late 1978 Paul Warnke, frustrated by the loss of too many arguments to Brzezinski, resigned from his posts as director of the ACDA and chief U.S. negotiator in several different arms control forums, including the CTB talks. George Seignious, a retired Army general, replaced him in his role as director, but not in any of the others.

Shortly after New Year 1979 Seignious telephoned to invite me to become the chief U.S. negotiator at the CTB talks.

I served in that post for two years, until the inauguration of President Reagan in 1981. During the first of those years, progress was held up by disputes internal to the negotiations. During the second, events external to the negotiations stopped us dead in our tracks.

The most important of the internal disputes involved the question of how many seismic monitoring stations would be deployed on British territory. We had all already agreed that the minimum number of stations needed to adequately monitor the USSR, by far the world's largest nation, was ten, and we had also agreed that the United States should also have ten on its territory. In the case of British territory, however, the American and British view, based on technical considerations, was that at most one station would suffice, while the Soviet view, based on notions of political parity, was that ten were needed there also. To that end, the Soviets drew up a list of proposed sites, one in Great Britain and the others on small British held islands all over the world. For a number of fairly good reasons, the British rejected all but the one at home, and the Americans supported them. This issue soon turned into an impasse that prevented us from making any headway on any other important issue.

A second major dispute involved how to handle the transition from a five year moratorium, which was, in effect, what we were then negotiating, to a permanent test ban, which was the stated purpose of the leaders of the negotiating states. Within the Washington decision-making apparatus, some agencies (ACDA and State) wanted to elaborate arrangements such that the actions to be taken at the end of the moratorium would automatically favor continuation, while others (DOE and the Military) wanted arrangements that would make continuation as difficult and unlikely as possible. We now know that a roughly parallel situation existed in the Soviet Union. Foreign Minister Gromyko and his supporters wanted a easy and automatic transition from a moratorium to a permanent ban and were able to sell a negotiation to that effect to the Politburo, while the Soviet military never agreed to anything more that a finite moratorium. Given this dispute within the home capitals, it is no surprise that we accomplished nothing in this regard in Geneva.

By the end of the first year much more serious problems arising in the world a large made further progress totally impossible, whether or not we might find solutions to the internal difficulties. These were first, the occupation of the American embassy in Teheran and the imprisonment of its personnel in November 1979, and then barely a month later, the invasion of Afghanistan by the Soviet Union.

The remainder of this essay discusses the "many causes of failure" in greater detail, and ends up with some general conclusions about them.

The Negotiating Team

Except for myself, the American CTB negotiating team consisted of persons carried over from Warnke's time. Happily for me, and for the good of the country, my old friend Gerald Johnson, who had been Warnke's deputy and Harold Brown's personal representative, agreed to stay in that position. Jerry was also a senior member of the SALT II negotiating team, the only person then serving on both delegations simultaneously.

The senior representative of ACDA (in theory, I represented the president, not the agency) was Alan Neidle, a professional Foreign Service officer and longtime tiller in the arms control vineyard. His zeal for success sometimes made him difficult to deal with, and the military and Energy Department representatives wanted him off the delegation. I decided I needed his experience, so I arranged for him to stay on for the next round.

General Edward Giller, USAF retired, represented the Joint Chiefs of Staff. Earlier, he had been the Energy Department representative on the interagency group that produced PRM-16. His bosses were against a test ban, and General Giller reflected their views accurately. This situation did, of course, carry the seeds of potential trouble, but Ed and I were old personal friends and had no difficulty in developing a good working relationship despite this fundamental difference. (When I was director of Livermore, Ed Giller was a young officer dealing with nuclear matters at the Air Force Special Weapons Center. We met then, and our careers crisscrossed in the ensuing years.)

Warren Heckrotte was one of several representatives of the Department of Energy. A fellow physics student at Berkeley many years before, he later joined the Livermore staff when I was director. While on the staff there, he became involved in nuclear test limitation questions as an adviser to the Department of Energy and the ACDA. In the early 1960s he served as a member of the U.S. team on the Eighteen Nation Disarmament Committee, and in the 1970s he was in Moscow when both the TTBT and the PNET were completed. All told, Warren has had more experience on negotiating teams than anyone else I know. He favored nuclear arms control in general but had some reservations about the practicality of a comprehensive ban on testing.

The British Delegation

Ambassador John Edmonds was head of the British delegation to the tripartite negotiations when I joined them.* Originally a naval officer, he had been able to transfer laterally to the Foreign Service when the British made drastic reductions in their navy after the Suez crisis of 1956. Later he also served as a diplomat at NATO headquarters, so he well understood the military and technical factors that called for nuclear tests as well as the political benefits to be gained from a ban on them. He was friendly, serious, pro-American, and typically English in his personal style, and we quickly developed a warm and lasting relationship. At first, Edmonds's deputy was Dennis Fakley, a nuclear weapons expert from the Ministry of Defense. Fakley was later replaced by Michael Warner, a senior civil service administrator in the same ministry.

The delegation also included several technical experts associated with the nuclear weapons establishment. Most were well known to the Americans who had been in similar work, and I had met several of them myself in that connection long before I had become involved in the negotiations.

---

*Sir Percy Cradock headed the British team when the negotiations started. At the end of 1977 he left to become ambassador in Beijing. Up to that time, John Edmonds had been his deputy.

The Soviet Delegation

Andronik Melkhonovitch Petrosyants was the head of the Soviet delegation.* The highest-ranking Armenian in the Soviet central government, he held the concurrent position back home of chairman of the State Committee for the Utilization of Nuclear Energy. His colleagues explained that his role in Moscow was equivalent to that of "an under secretary in your government" and that he preferred to be called chairman rather than ambassador. His responsibilities in this regard occasionally brought him into direct contact with the Soviet Politburo and with leaders of civil nuclear programs throughout the Soviet international system.

Short, balding, and already in his seventies, Chairman Petrosyants approached his job with considerable energy. He now and then showed a flare of temper, after which he would put on a wry smile and say something like "You know how it is; we southern peoples are very temperamental." (He still headed the civil nuclear program when the Chernobyl accident occurred, in 1986. I spotted him on a televised news conference that followed that disaster. He remained as feisty as he had been six years before.) He spoke only Russian, Armenian, and German. I could follow "diplomatic Russian" to a modest degree but could not make small talk in it, so he and I never had a conversation except through an interpreter. On rare occasions when we might meet without an interpreter present, I would comment on the weather in French, and he would do the same in German. That worked, but it probably didn't move the world forward very much.

Roland Makhmudovitch Timerbaev was the deputy head of the Soviet delegation. Half Russian and half Bashkir, he was a professional Foreign Service officer, dealing mainly with American and UN affairs, especially disarmament. His name was wonderfully contradictory. His mother had named him Roland after the medieval hero who kept the Moors out of France, his patronymic made him one of the sons of Muhammad, and the Timer in his surname came from the same root as that of Tamerlane, the scourge of Central Asia. He was well educated and spoke nearly perfect English. He knew a great deal about the facts and ideals of American politics and demonstrated a great curiosity to know even more. (As of this writing, he is deputy chief Soviet representative to the United Nations.)

Roland was an avid reader of the *International Herald Tribune,* and he occasionally quizzed me about the details of stories he read there. That newspaper also carried Art Buchwald's columns, and Timerbaev, like me,

---

*Morokhov was the first head of the Soviet delegation. At the same time he was also Petrosyants' deputy back in Moscow at the State Committee for the Utilization of Nuclear Energy.

had been a Buchwald fan for years. All I needed to do was say, "Remember the one about Sidney in Santo Domingo?" and we would both laugh heartily. Roland would then name one of his own favorites, and we would both laugh again.

Colonel Boris Tarasov was the third-ranking member of the Soviet delegation. He was from the Defense Ministry and spoke only Russian as far as I knew. He got along well enough with his opposite number in our delegation, Ed Giller, but I can scarcely recall his saying anything at informal gatherings, even when nearly everyone else managed to loosen up a bit.

The Soviet team also included a number of technical experts who came and went according to the topics under discussion. They were usually either from the State Committee for the Utilization of Nuclear Energy, the Ministry of Defense, or the Academy of Science's Institute on Geophysics.

There were also, I believe, at least two KGB men on the Soviet delegation in my day. One was, in keeping with the common practice, the secretary of the delegation, the man we would call if we wanted to make contact at a time not otherwise prearranged. The other was Peter Pogodin, their chief interpreter. Peter prided himself on being able to write what he thought was passable English doggerel. It usually was crude, sometimes racist and anti-Chinese. Years later I ran into Pogodin again, at a Pugwash meeting in Bjorkliden, Sweden, in 1984. He showed up there unexpectedly, in addition to Pavlichenko, the KGB man who normally had that duty.

Instructions from Home

The delegations in Geneva were all on very short leashes. We had some local maneuvering room with regard to tactics and scheduling, but in general Washington, London, and Moscow controlled the substance of our discussions tightly. I could, of course, try to influence the substance of our negotiating position, but I did so not by working things out with my colleagues in Geneva but by dealing with higher headquarters via telex and scrambler telephone. Indeed, I was really effective in this regard only when I was back in Washington, between rounds. Then I could meet directly with the two "backstopping" groups, one chaired by John Marcum of the White House staff, and the other by Admiral Thomas Davies, the ACDA's assistant director for multilateral negotiations. In addition to these "working level" committees, the Cabinet-level Special Consultative Committee (SCC) would meet when problems appropriate to it arose, typically once during each recess. If I was in Washington at the time, I would join that meeting.

The SCC meetings were usually attended by the top officials of the agencies involved. Most of the members fully supported the president's goals. Two of them—General David Jones, chairman of the JCS, and Secretary of Energy James Schlesinger—openly did not. They deemed the test ban a bad idea. Anything that worked to slow the negotiations or prevent or constrain their outcome was a step in the right direction. I did what I could to moderate this opposition.

On one memorable occasion I met with the chiefs as a body in their headquarters in the Pentagon. All of those at the table except me wore

four stars and so, it seemed, did their aides. I knew they were to a man opposed to the undertaking I represented. They were courteous, and I had met and come to know several of them before, but even so it was for me a tense time. Down deep I was a patriotic country boy. (My father, though never preachy about it, had repeatedly told me never to let the flag touch the ground.) I simply couldn't feel happy or relaxed face to face with our top military leaders while representing a course of action they all felt was not in our national interest. I had similar meetings with Energy Department officials, not including the secretary. In that case they were all people of a sort I was more accustomed to dealing with, and so their obvious hostility to my purpose was much less unsettling. In both instances I tried to explain why I thought the president's objective—to end all nuclear tests for all time—was in our national interest. I also looked for any modifications in our approach that might achieve nearly the same purpose but be less unacceptable to those with whom I was meeting. As in my earlier, similar attempt working with James Wade in OSD, I found no useful compromises.

As a result of these differences among the principals, the SCC itself often ended its meetings without being able to agree on a specific set of instructions to the delegation. When that happened, the various options and opinions would be taken by Brzezinski directly to the president. Carter would resolve them then and there on his own. Since Brzezinski himself was not really interested in a CTB, it is not surprising that this process often produced instructions I did not find particularly useful.

Locale and Schedule

The negotiations all took place in Geneva. Typically, the delegations would spend six weeks there, meeting with each other, and then return to their capitals for another six weeks. There we digested what had happened and worked with home officials on the development of instructions and tactics for the next round.

Our delegation made its headquarters on the Rue de Lausanne, a little over a mile from the place where the waters of Lake Geneva flow out into the Rhone. The building was called The Botanique, for a typically delightful and perfectly manicured Swiss garden park just across the street. The Geneva headquarters of the UN were only two blocks away, and many other world organizations were housed in the vicinity. The U.S. SALT delegation occupied the two floors above us; the U.S. delegation to the standing consultative commission and the U.S. ambassador to the international trade commissions were on floors below us. The ground floor was occupied by a lamp shop, which never seemed to have customers. Except for that, access to the building was controlled by Marine guards. Geneva might be a city dedicated to peace, but U.S. missions everywhere seemed to be fair game for terrorists. We could not be sure we were different.

Our Marines always wore their brightly colored uniforms inside our building, but not on the street, where doing so was prohibited by Swiss law. Their changing room was right next to my office, and occasionally they forgot where they were and became a little boisterous and noisy, but I was glad they were around looking after us, so I never objected.

The British delegation was housed in an office building nearby. The Soviet CTB delegation made its headquarters in the main Soviet mission to the UN, a large, walled-off compound near the Palais des Nations. When I first arrived, it was guarded by young KGB men dressed in civvies, but some months later they all suddenly sprouted dark brown uniforms like those the border guards wear all over the USSR itself.

Form and Style

The international meetings among the three delegations came in many forms and styles. The plenary sessions were the most formal. They occurred once or twice a week, depending on how fast things were moving. They were held on a rotating basis in each of the three missions, the host of the day serving as overall chairman. The entire professional membership of the three delegations, up to fifty persons in all, sat around a large table. Small placards in front of each head of delegation identified us, not by our names but by our countries: United States of America, Union of Soviet Socialist Republics, or United Kingdom. Only the three ambassadors spoke at the plenary sessions, always in their own language, and in short paragraphs followed by a sequential translation by the speaker's own interpreter.

The style and form of these plenary sessions served as a powerful reminder that we were not there on our own or to express our individual opinions. We were there purely as representatives of our various chiefs of government, and it was their opinions alone that we were presenting. It took a little getting used to, but it obviously had to be that way. (Years later I watched Henry Kissinger respond to a press question in which he was asked whether Anatoly Dobrynin, the longtime Soviet ambassador in Washington, had been honest and sincere. Kissinger said he did not know about that, but Dobrynin had certainly always been "accurate" and that was what mattered. Whatever he himself might have thought, he always fully reflected his bosses' views, and that was essential for effective state-to-state communication.)

The statements we made at the plenary sessions were, in general, read from papers that had been carefully prepared in a collaborative effort within each delegation. It was during these preparatory sessions that all members of the delegation (at least in those of the United States and the United Kingdom) had a chance to speak their minds and present the views of their home agencies. Once the statement was finally agreed, however, it became the sole responsibility of the ambassador to present it and explain it. Only the agreements made at the plenary sessions had any status in the negotiations. Details usually were, of necessity, elaborated in smaller working groups, but until the reports of these groups were accepted at a plenary, they had no formal standing.

When a plenary session ended, those present split into two groups. One group was made up of the "heads of delegation" and the other of the rest of those present. The remainder customarily stayed on in the main meeting room, where they gathered in small, informal subgroups to chat, drink tea or coffee, and eat cookies and peanuts. The "heads of delegation" group typically included the three highest-ranking persons in each delegation plus their interpreters, a total of twelve in all. This group retired to more comfortable quarters, typically a smaller room with a number of easy chairs and sofas arranged around a coffee table. On the table were decanters of coffee, tea, and orange juice and bottles of alcoholic beverages appropriate to the host mission. In our case that meant no liquor at all if the meeting was in the morning and both scotch and bourbon if it was later in the day. (When I arrived on the scene, no alcohol was served at the U.S. mission at any time of day, but I changed that under gentle chiding from the British. They explained that this was the reason the informal postplenary sessions were so much shorter when they were held at the U.S. mission.)

These "heads of delegation" meetings were more relaxed than the plenaries, but the first round of conversation nevertheless consisted of comments and questions by the ambassadors only. After the opening, however, others could speak and bring up topics they were especially concerned about. These smaller sessions provided an opportunity for further clarifying issues that had come up in the plenary. They were also typically used for elaborating the working calendar for the period immediately ahead—including, above all, the dates for adjourning the current round and opening the following one. Even in that case we followed a particular pattern. The Soviets often let us know by various indirect means what adjournment date they preferred, but they never formally proposed one. They always left that for the Western delegations. There must have been some bureaucratic reason for that ritual, but I never learned what it was.

Working groups also operated in a more relaxed way, partly because they were smaller and partly because they dealt with narrower questions. Ritual and formality still played a role, however. In particular, the ambassadors were, as a practical matter, forbidden to attend. On one occasion when I hinted I would like to be present to hear some of the arguments, it was made clear that the other ambassadors would then have to come, too, and that this would guarantee that nothing would be accomplished.

Less formal meetings of various other kinds also played an important role in moving the negotiations along. One type consisted of a carefully programmed sequence of trilateral lunches and receptions. The lunches were typically attended only by the "heads of delegations," as defined above. The receptions were for all of the members of all delegations, secretaries and other staff included. In addition, there was a carefully integrated rotating series of bilateral lunches.

The trilateral lunches might start out a little stiffly, but in time they loosened up. Alcohol was the main reason for the relaxation, but food and environment played a role as well. They usually went on for about two hours. Toward the end we inevitably raised toasts to success and the ladies, the latter in deference to a Russian custom. We toasted our wives, our mothers, and the female interpreters present. These references to women were never lascivious, just a bit maudlin.

There was also plenty of good humor on those occasions. I recall one bilateral lunch at the Soviet mission particularly. When it was just us "superpowers" off by ourselves, the Russians became even more relaxed. After the basic inhibitions of everyone present had been well washed away, the conversation somehow turned to the topic of Stalin's speeches. One of the Americans remarked that the formal records of these speeches, as published in the daily press and elsewhere, were always punctuated with frequent parenthetical remarks. These would range from a simple "Hurrah,

Hurrah!" on through "Standing ovation" and up to "Prolonged standing ovation, with thunderous applause" I became a little concerned with this choice of topic, but, to my surprise, everyone seemed to find the matter very funny. Finally, one of the Soviets remarked, "You forgot the highest level." After a moment pause he supplied it: "All rise and sing the 'Internationale.'" Everyone laughed at the preposterousness of it all, and then we went on to some simpler subject, perhaps another toast to our wives and mothers.

I am not a heavy drinker in real life, and I used all sorts of stratagems to avoid drinking more than I could easily handle. I thought I succeeded in general, but some of the photos the Russians took at these affairs—and then gave me—indicated otherwise (see the photo section for an example).

There were other, even smaller and more private meetings. These included many one-on-one meetings between John Edmonds and me. In them we could almost entirely ignore the long leashes that otherwise bound us and have really frank discussions about the problems we faced. I also had several useful two-on-two meetings with the Soviets. These always involved Petrosyants and Timerbaev (acting as both deputy and interpreter) on their side and usually just Jerry Johnson and me on our side. They invariably took place in Chairman Petrosyants's rather modest room at the Epsom Hotel and were always accompanied by the usual vodka and brandy with peanuts and potato chips. They were franker and freer than any other talks I ever had with the chairman, but a substantial gulf always remained. The language barrier and, even more, the political and historical differences between us were simply too great. Those factors did not, I believe, hamper the negotiations in any important way, but they did affect our personal relationship. To help make up for that, I was on a few very delightful occasions able to meet privately with Roland Timerbaev.

When Petrosyants was in Geneva, diplomatic protocol prevented private meetings between Roland and me. However, on a number of occasions the chairman had to be back in Moscow in connection with his job as head of civilian nuclear energy development; when he was, Roland became acting head of delegation and we could properly meet. The language barrier disappeared, and the even greater privacy helped overcome the other barriers as well. Even then, however, we both carefully avoided certain forbidden subjects—such as how many seismic stations the British might be persuaded to settle for—but we had good and fairly open discussions about other matters of mutual interest, including topics not directly connected to the CTB.

International Athletics

There was yet another and higher level of relaxation and informality: international athletics. One such event was a strictly U.S.–U.K. affair: an evening volleyball match in a local school gymnasium. To our surprise the British team included two men unknown to us but said to be from the local British chauffeurs pool. They added a lot of talent to the British team, but we won anyway. John Edmonds and I served as team captains, and we both stayed in there doing our best almost the entire time. It was great sport, but not enough to give any long-term aerobic benefit. Herbert Okun, my deputy at the time, marched up and down the sidelines with a reversible sign saying "Go Brits" on one side and "Go Yanks" on the other. Soon thereafter, but probably not for that reason, Herb was named ambassador to East Germany. In announcing the news to me, he said he was about to become an ambassador to a "real country," not merely to "initials," as I was.

Another memorable occasion involved all three delegations in a game of lawn bowling at John Edmonds's residence. We broke up into two teams, with Americans, British, and Soviets on each. The only difficulty we encountered was Peter Pogodin's penchant for stepping over the line when it was his turn to bowl. Happily for the good of the world at large, his Soviet teammates usually called him on it first.

The last athletic event I can recall took place at my residence in Coppet. It was during the later hours of an evening reception. The air was pleasantly warm, and many of the guests were on the patio overlooking Lac Leman. A solid young American naval officer and a burly Soviet technologist, both fairly far into their cups, were engaged in a very serious arm-wrestling match. Some others had already tested each other out, but those prior matches all ended easily and quickly one way or the other without incident. This contest stretched out with neither party winning. The men's faces turned deep red, and the air grew tense. Each of them had little tolerance for what the other represented. Each acted as though he absolutely had to win, and the rest of us, I included, became tense as well. To make matters worse, they were using a table with a glass top, and anything could have happened. I turned away for a moment, and it all suddenly ended. I don't remember who won, but they both walked away from the scene led by others in their own groups.

These special international contests aside, most of my physical activity consisted of hiking in the Alps and along the lake. Sybil and I took as much advantage of those glorious opportunities as time permitted, but the call of duty prevented me from doing anywhere near as much of it as I wished.

The Treaty and the Separate Verification Agreement

The Comprehensive Test Ban treaty was intended to be universal, and all states would be urged to accede to it. It would go into effect, however, as soon as the first twenty signatures were obtained, provided these included those of the United States, the United Kingdom, and the Soviet Union.

The main focus of attention of the tripartite negotiations, as well as of the negotiations within the governments in Washington, London, and Moscow, was less on the CTB treaty itself—that was seen as a relatively easy matter—than on an ancillary document, the so-called Separate Verification Agreement (SVA). This separate agreement would govern the way each of the negotiating powers satisfied itself that the other two were complying with the treaty. Each negotiating power had a long history of conducting tests, and each was much more concerned about compliance by its negotiating partners than by the world at large. (Of course, it was really the United States and the United Kingdom on one side concerned about the Soviets on the other, and vice versa, but the treaty and the SVA were drafted as though all three were independent actors.) Thus, although we always spoke of our negotiations as being about the CTB treaty, they were, in fact, very largely devoted to the separate, though closely coupled, SVA.

The three negotiating parties were very well aware there were two other overt nuclear powers in the world—France and China. We all knew that they would eventually have to become full partners in any comprehensive agreement to ban tests. For the time being, however, we tacitly agreed we could usefully go ahead without them. The two superpowers alone possessed over 97 percent of all nuclear weapons, conducted the great majority of all tests, and otherwise dominated the military scene. If those two, with the British, could work something out, then that treaty, or some modification of it, ought eventually to prove attractive to the others.

In any event, political realities dictated our going ahead without France and China. They made it clear they would not participate if invited. Moreover, the negotiations were already very complex and difficult with just three parties. Adding two more, each with its own very particular views, would have made the negotiations impossible. In sum, we three not only believed we could profitably go ahead without the other two; we also knew that doing so was both necessary and desirable. We were making a virtue of necessity, not an uncommon practice in the diplomatic world.

Substance

During the first two Carter years, the CTB negotiations moved forward rapidly. They came closer to a successful conclusion than ever before. By the time I became chief negotiator, in January 1979, virtually all the general agreements needed to complete the treaty had been worked out but only about half of the details. Persons with more experience than I thought we should be able to finish the whole process in six more months if we had sufficient "political will" in back of us. I agreed with that estimate.

By the end of 1978, general agreements covering four key issues had been achieved. First, national seismic stations (NSS), designed to detect and identify nuclear explosions, would be emplaced on the territory of each of the three negotiating powers. Second, there would be a provision for on-site inspections (OSIs) in case of ambiguous or suspicious events. Third, the treaty would run for a finite period, before the end of which another conference would be held to determine how and whether to continue it. Fourth, for at least the initial treaty period, nuclear explosions for "peaceful" or "economic purposes" would be outlawed, along with all explosions for the purpose of testing nuclear weapons. The main task still to be accomplished when I arrived on the scene was that of working out the remaining half of the details.

The national seismic stations, or NSS, were an old idea. Originally conceived back in the 1960s, they were then called black boxes. Their principal element was a seismometer that would be buried in the ground at carefully selected places on the territories of the contracting powers. The NSS would be operated by personnel of the host nation and provided with electrical power and communications facilities necessary to keep them going continuously and to transmit the data they produced, out to a central collecting point in a timely fashion. Each seismometer would also be provided with a special "authenticator"—in essence, a device based on cryptological principles that manipulated the data in such a way as to guarantee that they were unchanged in any way.

Early in 1978 a subcommittee to work out the technical characteristics of the seismometers was set up. The original American position concerning the manufacture of the NSS was that, once the technical characteristics had been agreed, it did not matter where they were actually made, provided they met those characteristics. We, of course, would supply the authenticating devices needed to assure the reliability of the data, but the rest of the device could be manufactured anywhere, even in Outer Mongolia, as one American diplomat once put it. Later, under pressure from the Joint Chiefs of Staff and the Department of Energy, Carter changed the U.S. position to insist that any NSS deployed in the USSR had to be entirely manufactured in the United States. The Soviets never absolutely rejected that change, but they never accepted it either. They simply said that was not how they saw it.

As a first step toward fixing the locations of the NSS, the parties on each side had handed over a list of the general approximate locations it wanted on the territories of the other.* In the next step, the receiving side would select a precise location within each limited general area. To the surprise of all parties, the question of the number and the location of NSS on British territory turned out to be by far the most difficult of the three cases. Although we did not anticipate it at the time, this issue proved to be the most intractable of the many internal problems faced by the negotiators.

There was a similar impasse concerning data transmission. The Western view was that seismic data must be made available very promptly, preferably in real time. We proposed satellite data relay as the best means for doing so. Indeed, given the remote locations of many of the sites, it was the only possible means. The Soviets expressed reservations about letting such data out of their country without the opportunity to examine them first. As they saw it, that probably excluded the possibility of satellite transmission, even within the USSR. The Soviets seemed to believe that we would somehow use this system to transmit illegally obtained data out of their country. I never could figure out how they thought we might do that, but something was obviously bothering them in this regard.

At the end of 1980, when we finally adjourned the Carter round of negotiations for good, most of these NSS issues were still unresolved.

On-site inspections had been a matter of contention between the Soviets and ourselves for many years. As far back as the Eisenhower administration, we had proposed them and the Soviets had rejected them. We wanted a specific annual quota of mandatory inspections. Except for a brief interval during the middle Khrushchev years, when they apparently were ready to agree to a very small quota, the Soviets always said no. In 1977–77 both sides made significant concessions in the area. We agreed to drop both "mandatory" and "quota" from our description of the idea. The Soviets agreed to accept the notion of "voluntary" OSIs in case of genuinely suspicious or ambiguous events.

The compromise involved a hierarchy of challenges and responses. For example, we would be required to explain the origin of our suspicions, and they would be required to reply directly and concretely to them. If we were not satisfied with their reply, we would request an OSI. If they then refused, we would have a prima facie case for accusing them of a violation, and we could take that accusation and the data to the UN Security Council, or to the world at large, and otherwise act as we saw fit. It may seem that there is a big difference between *mandatory* and *voluntary*

---

*Here, and in many other places, I recount the events as if the British and Americans made up one side and the Soviets another. Things did not always work that simply.

OSIs, but there really isn't. If the Soviets were to cheat, they would certainly never allow foreign inspectors to find the evidence, regardless of any agreements to the contrary.

At the time I took over the negotiations, the parties had agreed to establish the Joint Consultative Commission, or JCC, modeled after the Standing Consultative Commission established by SALT I. The JCC would handle all the questions arising at the time of a suspicious or ambiguous event and manage the arrangements for an OSI. Most of the details of an OSI had been worked out, but some questions remained, including the maximum delay that could occur before they undertook their work. More important, we had not been able to agree on the kind of information that could be used in support of a claim for an OSI. The Soviets were insisting on the use of seismic data only, and we wanted to be able to use more general information. (The original U.S. position, back in the 1960s, bad been that seismic data alone would trigger an OSI. When we accepted the notion of voluntary OSIs in 1977, we argued that the basis for asking for an OSI should no longer be so limited and that other physical evidence could be used as well. The Soviets seemed to fear that we might go so far as to base our requests on mere rumors.) Nor could we completely agree on the details of the tools and instruments the inspectors could bring with them. Most of these matters were still unresolved two years later.

Carter's original position on the duration of the treaty was that it should be forever. There would be, of course, the usual termination-for-cause clause, but there would be no allowance for periodic reconsideration. The Soviets, on the other hand, said the durations should be only three years. At the end of that time, we would reconsider the whole thing, and one of the factors that could influence whether or not it would be extended would be continued testing by others, meaning by France and China. The United States continued to insist on "forever" and on leaving out any references, however indirect, to France and China.

In the meantime, on the U.S. side, both the Joint Chiefs and the Energy Department also objected to "forever" and favored a finite, short duration. Bowing to the pressure on his right, Carter accepted the notion of finite duration but set it at five years. The Soviets were surprised but pleased. A few months later, again in response to internal pressures, Carter accepted the idea of three-year duration but continued to reject the idea that continued testing by others should be referred to explicitly in the treaty text.

In similar fashion many of the details relating to the connection between the Separate Verification Agreement and the CTB itself still remained to be resolved at the end of 1978, and so did such things as the precise wording of the preamble. Along with a number of lesser issues, most of these also remained unresolved when Reagan was elected president.

The Case of the British Seismic Stations

When, at the end of 1978, the CTB negotiators were considering the question of where the seismic stations should be deployed, everyone focused on the Soviet and American cases. Would the Soviets accept enough sites? Would they accept the locations we needed, or would they try to keep us out of certain sensitive but necessary areas? Would they pick sites in the United States on the basis of a genuine need or for some mischievous purpose?

No one anticipated problems with the British in this regard. Therefore, when on a gray, snowy day in January 1978, on which the men who should have been putting sand on the icy roads were out on strike, the British Cabinet finally got around to considering the Soviet proposal in detail and summarily rejected it, we were all caught flat-footed.[3]

The most serious problem the British had with the Soviet proposal concerned the locations. They made up, in retrospect, a very strange list:

Eskdalemuir, Scotland
Aldabra Is.
Brunei, Borneo
Tarawa Is.
Pitcairn Is.
Malden Is.
Port Stanley, Falkland Is.
Egmont Is., Chagos Archipelago
Belize, Br. Honduras
Hong Kong

The first location, Eskdalemuir, a seismically quiet site in the northern part of Great Britain, made sense. Almost all the other sites were impossible. They seemed to have been selected by someone in Moscow using an old atlas or perhaps a child's stamp album as a guide. They made little sense in the modern world.

Belize had been self-governing since 1964 and was about to become independent in 1981. Tarawa and Malden were to become constituent parts of the independent republic of Kiribati that very summer. The sultan of Brunei had long had total control over internal affairs, and by the end

of 1983 his country would be totally independent. None of these five locations would be on British territory at the time a treaty went into effect.

The Falklands and Hong Kong were worse. They could be expected to
remain British for some time, but in each case a neighboring power raised
conflicting claims of sovereignty. One can only imagine the reactions of
either the Argentines or the Chinese to the Russians' installing instruments in those places in order to detect nuclear explosions on the territory
of guess what adjacent country. In sum, seven out of ten sites were inap-
propriate, and five were even illegal.

I personally thought the Falkland Islands site made a lot of sense. Situ-
ated on the South American continental shelf, they made a good location
for monitoring any nuclear tests in Argentina and Brazil, both of which
had pointedly refused to ratify the Nonproliferation Treaty. The British government, however, was already anticipating the sort of blowup that actually
happened only a couple of years later. It was totally unwilling to discuss anything touching on the Falklands—no matter how noble the purpose—with
anybody, not even with its close allies, and certainly not with the Russians.

The selection of Pitcairn Island posed a special problem all its own.
There were no good port facilities on the island, so getting a complicated
device like an NSS onshore and installed promised to be a formidable
task. At one point later in the negotiations, Ambassador Edmonds and I
tried to joke with Chairman Petrosyants about the problem. The occasion
was a reception at my residence, and the three of us were joined only by
Peter Pogodin, acting as interpreter. Edmonds went on at length with
plenty of hand gestures, explaining how the three of us would have to
wade ashore through the surf carrying a this apparatus in our bare hands.
Petrosyants looked more and more perplexed. Finally he turned to Pogodin and said, "What is this *Pitcairn* they are talking about?"

After that question was duly translated, Edmonds expostulated. "It is
one of the islands *you* selected for an NSS!" Petrosyants snorted and
changed the subject.

But a highly peculiar set of locations was not the only problem. Cost
was another. We still did not know what the price of an individual NSS
would be, but a few million dollars seemed likely. When multiplied by
ten that came to a pretty penny by British standards. Up to this point no
one seemed to have thought much about that issue. The Foreign and
Commonwealth Office, which managed the negotiations, was not accustomed to spending that kind of sum on that sort of project. It turned
naturally to the Ministry of Defense. There the problem was intercepted by Victor Macklen, an artful bureaucrat who used it to bring
further internal consideration of the issue to a near standstill.

Macklen had long been an important figure in British nuclear circles. At the time he was deputy chief scientist in the Ministry of Defense, and thus Dennis Fakley's immediate boss. Adamantly opposed to a test ban himself, he was in close touch with those Americans who also were, including General Giller and Donald Kerr. Kerr and Macklen were, in effect, part of a larger group of insiders dedicated to thwarting Carter's desire to negotiate a test ban. Others especially effective in this regard were Admiral Bob Monroe, the chief of the Defense Nuclear Agency; Julio Torres, one of Kerr's assistants; and the notorious Richard Perle, then still on Senator Henry Jackson's staff. Each was an experienced bureaucratic infighter; all were determined to stop any progress toward a test ban, no matter what their top national leaders might think or want. In this instance Macklen, with whatever help his American coconspirators were able to give him, used the NSS issue to develop opposition to the nine remote NSS within British defense circles. There the possibility of having to spend even a farthing on them was especially unwelcome. This opposition evidently played a role in persuading Prime Minister Callaghan on that gray January day to reject the Soviet list of sites.

This turn of events took all the delegations by surprise. I, of course, had just arrived on the scene of the negotiations themselves, so everything was new and, often, surprising to me. This development, quite obviously, amazed my experienced colleagues as well.

The Soviets, when we informed them a few weeks later, were even more astonished. Once they had caught their collective breath, they flatly rejected the British proposal that there be only one NSS on their territory. We supported the British position by pointing out what a small country the United Kingdom is, but the Soviets would have none of it. This is a political issue, they said, not a technical one, and so it must be resolved on the basis of an appropriate political principle. In this case that principle was clearly the principle of equality. Furthermore, in their technical judgment, no NSS were needed anywhere. The current deployment of seismographs all over the world was adequate for the job and did not need to be augmented. They understood, however, that the U.S. Senate would never ratify a treaty without NSS on Soviet territory, and so they were willing—for that basically political reason—to accept NSS as a condition for a CTB. In sum, they said, either there are ten stations in the USSR, ten in the United States, and ten in the United Kingdom or it's 1-1-1, or 0-0-0, but never, no never, 10-10-1.

To emphasize the seriousness of the issue, the Soviets went on to say there was no point in holding further discussions on the technical characteristics of the NSS until we could settle the matter of their number.

As soon as we understood just how serious the number issue was, we started to look for a suitable compromise. Perhaps if the British could be persuaded to come up to some other number—not ten, but more than one—the Soviets would come off their 10-10-10 stance.

But before we could even begin to work out a compromise, an additional complication intervened. The political situation in London made it necessary for Callaghan to call for new elections. Very quickly, I knew, London would become totally absorbed in these events. If anything was to be done, it had to be done fast. I decided to do something slightly out of line for a CTB ambassador. Instead of going through the ACDA bureaucracy, as was the norm, I sent an encrypted telegram explaining the situation directly to Secretary of State Vance. (All of the hundreds of daily telegrams from U.S. ambassadors to their home offices are addressed to the secretary of state, but he does not normally receive or read them unless certain extra words are added to their routing instructions. In this case, I used those extra words.)

I suggested that Vance intervene personally in the matter. He seemed willing, but the idea raised a storm of objections at both ideological extremes. The ACDA said, no, we should insist that Britain accept ten, just like the big boys. The Joint Chiefs of Staff and the Department of Energy said, no, we should not suggest a compromise, because the British are right. It also quickly became clear that any suggestion that the United States supply the British with NSS free of charge would be fought tooth and nail by those who wanted no treaty at all.

This conjunction of arguments created an impasse within the U.S. government that was too hot for the working-level backstopping group to handle. The issue had to be considered by the Cabinet-level SCC before Vance could intervene with his British counterparts. That process would take much more time, and before it could happen, Britain had a new government, with Margaret Thatcher at its head.

Prime Minister Thatcher had been educated as a chemist. Like Jimmy Carter, who had a background in nuclear engineering, she felt she could personally understand the technical details of the test ban issue. On looking into them, she decided she was against it. Even before she took office, some opponents of the treaty, probably Victor Macklen and perhaps one of the American opponents as well, met with her and presented their side of the argument, which she accepted. Later, when the American and British negotiators finally had a chance formally to propose that she accept more than one NSS on British territory, her response was, in effect, "If one was good enough for Callaghan, one is certainly good enough for me." In essence, her government continued to follow the U.S. lead and to support the U.S. position that a CTB was in the Western interest, even though she personally had reservations about it. Whenever difficulties arose in the negotiations, however, she did nothing to help overcome them.

(In 1985, long after the end of the Carter round of negotiations, I had a chance to discuss this whole episode individually with Jimmy Carter and David Owen, the British foreign minister in the Callaghan government. They both firmly recollected that a compromise in which three NSS would be placed on British territory had been worked out. I was very surprised by that claim and said so. However, each man independently stuck to his story, though neither could recall further details. It seems possible, therefore, that in some completely private venue Vance and Carter had agreed to that idea, and then presented it to Owen, who also accepted it. But whatever may have been privately discussed and tentatively arranged, it had no official status whatsoever. Neither Edmonds nor I, despite much urging and prying, ever got a number from either of our headquarters different from one, even on a very informal or confidential basis.)

At the same time when I was trying to persuade the British to accept some number of NSS greater than one, I was doing my best to persuade the Soviets that one was just right. On more than one occasion I noted that the Soviet Union was a very large country and that at least ten seismic stations would be needed to monitor events on its territory. The United States is big, too, so ten stations are justified there as well. But the United Kingdom is a comparatively small country, for which one station surely suffices.

Petrosyants generally stuck to his argument that the matter was purely political in nature and that my technical arguments were therefore irrelevant. Once, however, he replied directly to them.

"Da," he said proudly, holding his hands out in a wide gesture confirming great size, *"Sovietskii Soyuz ogromnaya strana*—Yes, the Soviet Union is a huge country."

Holding his hands somewhat closer together, he went on to say, "And, the United States, too, is a big country."

Then he gestured even more broadly with both hands and waved his ten fingers in all directions, adding, "But the British Empire is all over the world."

To personalize the whole affair, he said he was less concerned about Ambassador Edmonds's testing his own bomb in some remote place, than about Edmonds's allowing Ambassador York to test one of his on British territory. That happened in the past, he noted, so why not again?

The matter of the British NSS remained unresolved for the rest of the negotiations. The British position continued to be that one was enough. The American position, as presented to the Soviets, was that the British position made good sense. The American position, as presented to the British, was that they ought to accept some small number of NSS on overseas territories. The Soviet position was that the British and American positions were unacceptable and that they could not sensibly discuss anything else about seismic stations until that basic question was resolved.

Other NSS Issues

During the next two years the Soviets made two exceptions to their refusal to discuss the technical characteristics of the NSS. One instance involved a visit by Soviet specialists to the Sandia Laboratory and other U.S. institutions where our version of the NSS was being developed.

The invitation to the Soviets had originally been made in the fall of 1978, when Paul Warnke was still ambassador. When the Soviets finally accepted it—some six months later, on my watch—I expressed my pleasure. I immediately informed Washington, expecting the authorities there to proceed promptly with all the necessary arrangements. To my horror, Washington wired back instructing me, in effect, to hold on, to avoid making any final agreements about the visit, even regarding dates, until certain problems could be worked out. Bureaucratic opponents of a test ban had raised several eleventh-hour issues—including those of security and funding—and until these were resolved I was to do nothing more. The opponents were not concerned about the issues themselves; they simply wanted to delay or, if possible, stop the whole process.

I found myself in an extremely awkward position. On the one hand, I was telling the Soviets how pleased the United States was that they had finally accepted its invitation. On the other, I was declining to agree to any specific arrangements, even to the date we would accept them. "We are still working out the mechanics," I said on more than one occasion. "I will let you know about them as soon as I can."

I felt as precarious as a tightrope walker, but somehow we managed to bring off the visit in the summer of 1979. As Jerry Johnson often remarked, the most difficult times were when the Soviets unexpectedly said yes.

The other exception to their refusal to talk about technical characteristics came in 1980. To our surprise they suddenly agreed to discuss those of the down-hole components of the seismometers themselves, but still not those of any of the support equipment, including that needed to record or communicate the data. Additional experts from all three countries temporarily joined the negotiations. The academician Mikhail A. Sadovsky, a very distinguished Russian scientist and head of the Soviet Institute of Geophysics, was among them. In just a few weeks we were able to work things out to everyone's satisfaction. The agreement on those technical characteristics turned out to be the last substantive thing we accomplished.

Stopping Over in London...

Every time Sybil and I went to Geneva to start a new round, we stopped over in London. We did the same on our way back home six or so weeks later. Happily, Rachel and John were living in London at the time. They had a house in Islington, which, though typically small and narrow, had an extra room just right for us.

At least once during each round trip, I checked in at the Foreign and Health Office in Whitehall. Edmonds and his equivalent of a backstopping group were headquartered there. We used the opportunity to make certain the two Western parties understood each other's positions and purposes, especially on matters on which they differed somewhat.

I also occasionally met with officials in the Ministry of Defense, including Victor Macklen. The people there were obviously troubled—to put it mildly—by the whole affair, and I wanted to understand their position as well as I could. On one stopover General Giller exercised his nuclear connections and arranged for several of us Americans to visit Aldermaston, the site of the AWRE, the British Atomic Weapons Research Establishment. We were able to have a good discussion about the British experience during past moratoriums (including a long self-imposed one) and to learn more about their hopes and plans for future tests at the Nevada Test Site. The close cooperation between the American and the British nuclear establishments was manifest. As I saw it, it benefited both sides—but, of course, especially the junior partner, Britain.

That relationship had its origins in the Manhattan Project itself. After the war it fell off for a time to virtually nothing, and the two parties proceeded along separate ways. In the 1950s cooperation was reestablished, gradually and tentatively at first. In May 1957 I was invited to be one of two U.S. representatives to witness the first British thermonuclear test, staged out of Christmas Island and conducted over Malden Island, in the mid-Pacific. I estimated the yield from the fireball size and found it relatively modest. I wondered about the design, but my hosts were not allowed to tell me anything about it. Now in 1979, more than twenty years later, the relationship was more intimate. In an utterly different context, one presumably devoted to stopping nuclear tests rather than to promoting them, I finally got a close look at the details of the bomb itself. My yield estimate had been correct, but the design turned out to be more advanced than I had guessed.

Much of the work at Aldermaston at the time was devoted to the Chevaline program, whose purpose was to provide a suitable warhead for the British Poseidon submarines. Considered in the context of British military preparedness only, the people at the Ministry of Defense and the AWRE did indeed need more tests to achieve their objectives. Considered in the larger context of international security as a whole, the world—including Britain—needed to move away from its heavy dependence on nuclear weapons. From the point of view of those in the nuclear establishment, the first set of considerations was compelling. From that of my new friends and colleagues in the Foreign Office, the second set was compelling. It was a microcosm of the situation in Washington, with only irrelevant differences in the details.

My frequent trips to London also allowed me to renew my close working relationship with Lord Solly Zuckerman. Solly, originally a zoologist, had become a major figure in British defense affairs during World War II. When I was working in the White House and Defense Department in the late 1950s and early 1960s, Solly served as chief science adviser to the Ministry of Defense; later he was adviser to the Cabinet as a whole. At that earlier time we saw each other often on official business, usually in Washington. Now, at the time of the CTB negotiations, he was in his seventies, formally retired from his principal government and university posts and living in Burnham Thorpe, in the northern part of East Anglia. Despite his formal retirement, he remained very active in London affairs. Not only did he participate in debates and other activities in the House of Lords, but, more important, he also maintained a place in the Cabinet Office in Whitehall. He strongly supported the push for a test ban and did what he could to overcome the opposition at No. 10 Downing Street and in the Ministry of Defense. To that end, he arranged for Sir Robert Armstrong, then newly appointed secretary of the Cabinet, to spend a weekend with me at Burnham Thorpe. Our joint purpose was to bring Sir Robert up to date on the negotiations and to make sure he fully understood the official American view of them. It was typical of Solly's approach to such problems—and, I believe, a successful and useful weekend.

Solly also was long (1955–84) head of the London zoo, a position he never fully relinquished, no matter what else he was doing. I took advantage of this side of him, too, and arranged for Cynthia, then a veterinary student at UC Davis, to spend two weeks working behind the scenes at the zoo. Solly was tied up when she first arrived, and planned to meet her at lunch toward the end of her stay. On the scheduled day Lord Mountbatten, last viceroy of India and Solly's longtime friend and close colleague, was assassinated by Irish terrorists. Cynthia lost that opportunity to meet Solly, and another never came. The long arm of terrorism reaches out and touches people in unexpected ways.

...and in Moscow

I wanted to visit Moscow, too, to see if there was anything I could do there to help clear the air and move the negotiations forward. That, however, was not as casually or as easily arranged as a stopover in London. I had to have a formal invitation. To get one, I regularly reported my visits to London to the Soviets. I boasted about how useful they were in moving things along and hinted that a visit to Moscow might do the same. In the summer of 1979, after several such hints, they extended an invitation to me for a visit by three U.S. officials and their wives. In June, Jerry and Mary Kay Johnson, Sybil and I, and John Marcum, my anchorman on the NSC staff, arrived in Moscow.

We first visited various elements of the State Committee for the Utilization of Nuclear Energy—Petrosyants's Moscow base—and the Geophysical Institute. That was very interesting, but not particularly relevant to our negotiations. We then made a side trip out to Obninsk, where we visited the "first-in-the-world" nuclear power plant and the underground seismic laboratory. There they showed us some of the seismic detection technology they had in mind for monitoring a test ban. The equipment was heavier and otherwise more crudely built than ours, but it did seem to measure seismic signals adequately, or nearly so.

The most important part of the trip was a brief visit with Georgi Kornienko, first deputy foreign minister and longtime expert on American affairs. He spoke excellent English, as did most of the other Soviets present, so no interpreters were needed. The meeting, however, did not go as I had hoped. Another of those pieces of bad luck in timing that have plagued the CTB for decades intervened.

Just weeks before our arrival in Moscow, the United States had detected an explosion at Semipalatinsk, in central Siberia, that appeared substantially to exceed the limits set by the Threshold Test Ban Treaty. That treaty was, unhappily, still unratified by the United States, but both sides had pledged to live within its restrictions. There were large uncertainties in our estimates of yields, and it was entirely possible that this recent one had not exceeded the limit, but it seemed to. At the last minute we received instructions to make the matter of this apparent violation the first order of business. We therefore opened our discussions by asking for more information about it. Kornienko smiled coldly and flatly denied they had exceeded the threshold then or at any other time.

He then added something we already knew. If we wanted to improve our ability to measure the power of such explosions, we could easily do so by ratifying the Threshold Test Ban Treaty. When we ratified, they would supply the additional geophysical data the treaty called for, and then perhaps we would be less prone to overstate the size of their explosions. We continued to speak past each other for a few more minutes, but to no result except to spoil whatever chance this visit might have provided for helping the CTB negotiations. In retrospect, I do not believe we could have accomplished very much in any event, but you never know, and it's always worth a try. We—or maybe the Russians— muffed that one.

Last Days in Geneva

As November 1980 approached, all of us, British and Soviets included, focused more and more attention on the forthcoming American presidential elections. Usually the incumbent—in this case Jimmy Carter—has a distinct advantage. This time, however, the negative impact of the ongoing crisis in Tehran began to make it look as though the challenger, Ronald Reagan, had a good chance to win.

In late October, a few weeks before the actual election, we obtained a videotape of the Reagan–Carter debate and arranged a private showing at the main U.S. mission to the UN offices in Geneva. Most of the British and the English-speaking Russians attended. They were not happy with what they saw. Obviously, Carter supported our efforts and could be expected to continue to do so. Almost as obviously, Reagan did not. Those who understood U.S. politics well—Edmonds and Timerbaev, in particular—felt that Reagan had won the debate and probably also the presidency. The future of the negotiations was in doubt, to say the least.

The elections were scheduled for November 5, just two days before the celebration of the anniversary of the Great October Revolution. The Russians wanted to be back in Moscow for that occasion. It is, I learned, a family holiday as well as a patriotic celebration, and it is accompanied by much partying and drinking. The Soviets therefore proposed that we adjourn the present round early enough for them to get back to Moscow for the festivities, and that meant before the U.S. election. I disagreed, saying we simply had to know how the U.S. election came out before we left Geneva. The Soviets put forth all sorts of arguments, including a plea that the academician Sadovsky, then a temporary member of the delegation, had to get back for his seventy-fifth birthday celebration with his family. Petrosyants also suggested we prepare two separate contingency plans for the future, one for each possible president. I insisted we needed to know the final outcome in order to focus our minds properly.

Petrosyants then made a Solomonic decision. Half the delegation would go home for the big party, himself and Sadovsky included, and the other half would stay in Geneva with Timerbaev acting as head. That satisfied me, so we agreed and set adjournment for November 12.

The election made us all lame ducks, Soviets and British as well as Americans. Neither Edmonds nor Timerbaev wanted to believe it was over. I knew it was, certainly for me personally and very probably for all of us.

I had a number of private meetings during those last days with John and Roland. Roland said these negotiations may well have been our last chance for making a test ban treaty that would effectively reinforce the nonproliferation regime and thus prevent the nuclear genie from getting totally out of control. He was obviously sincere. I told him I agreed with him.

We had one last plenary, at which I announced my final instructions from Washington. I was to fix no date for resumption. That would be determined later through diplomatic channels.

Later never came.

The Many Causes of Failure

In retrospect it is clear that the failure of the Carter round of CT negotiations was very overdetermined. Several major problems arose any one of which would have assured failure by itself. Certain groupings of lesser problems could have had the same effect. All of the main problems were external to the negotiations; most of the lesser ones were intrinsic to them.

The intrinsic problems arose both at the international level—that is, in the form of differences between the negotiating parties—and at the national level, as differences within the various national bureaucracies. Among the most important problems at the international level were the matter of the number of seismic stations, differences over both the form and the content of on-site inspections, and arguments about the treaty review process and the eventual role of France and China in it. All of these problems could probably have been resolved eventually, but failure to do so for any of them would have resulted in no treaty.

In Washington, at the national level, the most important of the intrinsic problems was the continuing total opposition to any CTB by powerful elements in the military and in the upper reaches of the Department of Energy. In addition, real differences over important details among those who favored a test ban, or who were at least tolerant of one, hampered progress throughout Carter's presidency.

In London important elements of the Ministry of Defense and, after May 1979, the new prime minister, Margaret Thatcher, took a dim view of the whole idea. The arguments over the numbers of NSS and how they would be paid for only made things worse.

There may have been important unresolved differences in Moscow also. Soviet reluctance to accept certain of our requirements for on-site inspections seemed to reflect one such difference. The Politburo, led by Brezhnev, had evidently decided that OSIs were both necessary and tolerable. That historic breakthrough was surely not easily accepted by the secret police. As usual, direct evidence of such a difference was hidden very effectively, but it cannot be that the KGB officers threw their hats in the air and cheered when they received the news about the Politburo's decision. The best evidence of this was the repeatedly expressed concern that our hidden purpose was somehow to obtain what they always called "information not necessary for purposes of verification." In any event, even though the Soviet government had accepted OSI in principle at the start of the Carter round, Soviet negotiators continued to argue over essential details until the very end. As always, we wanted easier, quicker, and surer access, and they wanted more restrictions on what we could do.

It is probable that with sufficient "political will" in the three capitals, these internal problems could have been overcome eventually. But before that proposition could be tested, three major external events intervened. Any one by itself would have eroded the available political will below the level needed for success. With the occurrence of all three in quick succession, the chance for a successful outcome dropped to zero.

The first to appear was a collection of disputes and delays experienced in the SALT II negotiations being conducted at the same time. Some of these were matters intrinsic to SALT itself; others were the result of external events. The most important intrinsic arguments were those over Soviet data encryption and the Soviet Backfire bomber. While these were still on the table, Deng Xiaoping, the Chinese leader, paid a state visit to Washington. The visit, which seemed to portend further improvements in U.S.–China relations, annoyed the Soviets. As long as news of Deng's visit was still reverberating in the world at large, the Soviets stalled in Geneva. As a result, when SALT II was finally signed in Vienna on June 18, 1979, it was already more than half a year behind schedule.

Then, in August, just as Senate consideration of the treaty was about to begin, Senator Frank Church suddenly raised the issue of a supposedly "new" battalion of Soviet troops in Cuba. It took months for Washington to discover not only that the battalion had been there all along but also that President Kennedy had known of it and implicitly agreed to its presence. Before this tangle of errors could be unraveled, two more disasters occurred.

The first was the capture of the American embassy in Tehran by a band of "student" followers of Ayatollah Khomeini. The ayatollah had recently become the only real power in Iran, and no one in Washington knew how to deal with him. As the weeks and months wore on with no resolution in sight, the image of the helpless giant being tormented by the Lilliputians took hold. Any action anywhere that could possibly be characterized as weak—or even accommodating—became unacceptable to more and larger segments of the U.S. body politic. Negotiating with the Soviets, even in the absence of information linking events in Tehran to Moscow, took on that color.

Important as this political link between events in Tehran and negotiations in Geneva may have been, the mechanical linkage was more so. After the capture of the embassy, it became a practical impossibility to get either the president—or any Cabinet-level group, including the SCC, to focus on the CTB negotiations or any other such side issue. The bureaucratic opponents of the CTB were quick to take advantage of that situation. They more and more often claimed that even minor questions required Cabinet-level consideration. Since such consideration could not be obtained, we American negotiators began to find it ever more difficult to respond to even the smallest propositions or questions our negotiating partners raised. Not only the political will to negotiate but even the ability to do so, in the mechanical sense, was destroyed by the events in Tehran. As if all that had not been enough, the Soviets invaded Afghanistan, only seven weeks later. That disaster happened during our Christmas–New Year recess. Before we went back to Geneva, the White House instructed us to proceed very slowly with the negotiations and not to engage in any overt social interactions with our Soviet counterparts. In addition, I was ordered to open the next round of negotiations with a formal condemnation of the invasion before proceeding with the substance of our work.

The British ambassador had parallel instructions. Since the same thing was happening in all other U.S.-Soviet diplomatic contacts, the Soviet CTB delegation was expecting it. They listened stone faced and said very little in reply, and we all then went on—albeit very slowly—with the main business at hand.

Other reactions by Carter also affected our negotiations, although not so directly. For one thing, he ordered tighter controls on the transfer of high technology to the Soviets. U.S. opponents of the CTB asserted that these expanded controls covered certain components of the seismic stations, and they tried to achieve yet more delay on the basis of that claim. More surprising, even the cancellation of the Olympics entered into our negotiations. During one of our semiformal "heads of delegation" meetings, Petrosyants suddenly proposed a toast to the Olympic Games. As an undiplomatic act, it deserved some sort of gold medal itself. God only knows what inspired him. Ambassador Edmonds, I, and the other Westerners present sat with our arms folded and scowled. All the Russians raised their glasses. Except for their leader, all of them were obviously sorely embarrassed by this unexpected turn of events. As they normally did in such cases, they stared with great determination at the floor or straight ahead at the opposite wall, carefully avoiding eye contact with any of us. I made some strongly negative comment, which I cannot remember, and we adjourned that meeting immediately thereafter.

Any one of these major developments alone would have doomed the CTB negotiations. It took some time for the seriousness of the problem with SALT II to become evident. When the embassy was first captured, we again could not know how long that miserable event would drag on. There thus remained, for a while, hope that the problems confronting us, external as well as internal, would be resolved in due time. But after the Soviet seizure of Kabul, it was obvious that all was lost.

I had to reflect on my own position in all this. Was there a point to going on, even in the face of such odds? Winston Churchill's one-liner "Jaw, jaw, jaw is better than waw, waw, waw" came to mind. Our government and, evidently, those of the United Kingdom and the USSR had clearly decided not to cut off all contacts. In particular, they intended to go on with the kind of exercise that engaged us. I easily concluded that continuing with the talks was clearly worth my personal time and effort. Talking is indeed better than many of the alternatives.

We took the restriction on social relations seriously. We no longer invited the Soviets to evening receptions, though we did maintain the custom of a rotating series of working lunches. Similarly, we declined invitations to come to their embassy to celebrate their various patriotic holidays.

I made one exception to the rule on socializing. When Timerbaev's wife, Nina, came to visit Geneva for a few weeks, I arranged for her and Sybil to have a quiet lunch at our residence in nearby Coppet. During my two years at the negotiations, she was the only Soviet wife who came to Switzerland. Most of the Americans, and many of the British, had their wives with them all of the time, but not the Soviets. Nina's visit, then, was a special and unusual occasion. In addition, she was a delightful person, with lots of verve and sparkle.

Sybil and Nina even talked about the Olympics, but not in the negative and provocative style of Petrosyants's toast. Nina and her working group back in Moscow had put in much overtime on peripheral activities intended to help make the games a success, and she had been looking forward to them as a bright and hopeful event. The withdrawal of the Americans was a major blow to those plans and hopes, and she was personally sad that it had happened. She did not, however, comment on whose fault it might have been.

What of the Future?

The most decisive of the factors causing the failure of the Carter round of CTB negotiations were all accidents peculiar to that time. If they had not happened, how would our negotiations have come out?

I am certain that with an adequate level of political will on the part of both Carter and Brezhnev, and despite the doubts of Thatcher, we could have achieved a treaty satisfactory to all three parties. Carter would have had to put down some of the more extreme demands of internal opponents of a CTB, including those relating to the treaty review process and the real-time reception of NSS data. Brezhnev would have had to accept basically our version of the on-site inspections. And both Brezhnev and Thatcher would have had to admit that there were numbers between one and ten, such as three. It would have required some hard work and some compromises, but it could have been done.

But what then? Would the treaty have been ratified? Aye, there's the rub! The Founding Fathers of our Republic were determined to avoid "entangling alliances," and they created machinery designed to assure just that. The most important and enduring element of this machinery is the requirement that treaty ratification be approved by a two-thirds vote of the Senate, the less representative of our two legislative bodies. That peculiarly American requirement kept us out of the League of Nations in 1919 and for fifty years blocked ratification of the 1925 convention outlawing chemical warfare. More generally, it has caused presidents to hesitate and lose opportunities on other occasions. It would, I fear, have blocked ratification of a CTB even if we had been able to elaborate one agreeable to the president in 1979 or 1980, and even if none of the external impediments had existed.

In retrospect my thirty years' experience with this issue tells me that probably at no time since the cold war started would the Joint Chiefs of Staff have endorsed—or even quietly acquiesced in—a comprehensive test ban, and, given that, no Senate would have assented to ratification.

The JCS and their allies in the nuclear community have used a number of arguments in opposing a CTB. Foremost among them are doubts about verification and the claimed need for tests to assure the continuing reliability of our stockpiled weapons. They believe what they say about those issues, but their real opposition is even more fundamental. In their view it simply makes no sense to halt weapons testing in a world where nuclear weapons are, and promise to remain, the cornerstone of our security policy. They are willing to accept a prohibition on the kind of tests that cause direct harm, such as tests in the atmosphere, but they see no sense in limiting tests that do not. They are, of course, strong supporters of our nonproliferation policy, but they do not agree with the claim that a CTB is essential to limiting further proliferation. The Joint Chiefs and the nuclear establishment do, in general, support arms control agreements that limit or modestly reduce the numbers of weapons or that restrict weapons altogether in certain areas, such as nuclear free zones. They will not, however, willingly endorse or acquiesce in restrictions on the tests needed to maintain readiness or to support the kind of modernization that is always under way in the military establishment.

In sum, however desirable a CTB may be, it seems not to be a promising option under current world conditions. Moreover, if another president were again to push hard for a CTB, doing so would, as it did in Carter's time, make it much more difficult for him to achieve other and, I think, more valuable forms of arms control, such as that involved in the SALT and the START negotiations.

After thirty years of actively working for a CTB, I have again been forced to conclude that one will be politically possible and stable only in a world in which the great powers are clearly and forcefully moving away from their current dependence on nuclear weapons. In a world in which strategic arms limitations and other nuclear restrictions were the order of the day, a CTB would not only be possible but even strongly reinforce those other actions. In a world like the one we currently have. however, a world without that precondition, a freestanding CTB is apparently not viable. (Nevertheless, I believe many other forms of arms control are currently possible, given a properly inspired and competent administration. I specifically include SALT-like limitations and reductions, the elimination of battlefield nuclear weapons, nuclear free zones in sensitive places, and the like. A CTB, uniquely among mainstream arms control proposals, draws an especially hostile response.)

In 1979 when President Carter invited me to be chief negotiator, I set such thoughts aside. The excitement of the opportunity and the hope that he could achieve controls on arms across a broad front combined to overcome my doubts, and I marched forth full of optimism. I would do it all over again under similar circumstances. Someday someone must succeed.

Wasn't it frustrating?

Friends back home frequently ask that question. The answer to it is short and simple. Yes, it was, but working things out in the Washington bureaucracy was much more frustrating than dealing with the Russians.

Are they serious?

It is seldom easy to decide exactly when the Soviets are serious and when they are engaged in political posturing. Even so, in the case of the CTB, I come down solidly on the side of concluding they were serious, certainly during the Carter years and most likely at other times as well. My long personal contacts with Kapitza, Emelyanov, Timerbaev, and Petrosyants and other high-ranking Soviets who dealt regularly and at length with the CTB and related subjects convinced me that these men, at least, were honestly in favor of stopping all nuclear tests, those of the USSR included. All of them had solid contacts with still higher authorities, sometimes even with members of the Politburo itself, and they therefore knew how the people at the top felt about the question. I believe the men I met would not have worked so hard and so long on the issue if they had not believed they had the backing of the top leaders.

That said, I must add that I do not know why the Politburo favored a CTB. Perhaps it was because its members were more concerned than we were about proliferation. They saw—correctly, I believe—a CTB as being a good as well as a necessary means for inhibiting proliferation. Perhaps, too, they felt they could not win a wide-open technological race with us in this field. Simply stopping the whole thing may have seemed better than allowing us to lead them and the rest of the world they knew not where.

But can we trust them?

The simple answer to that persistent question is that we don't. The entire negotiating process is designed to create a situation in which the need for trust is replaced by adequate verification policies and procedures. These may depend on a certain amount of overt cooperation, but not on unsupported trust.

More basically, I hold that *trust* is a word that should be used only to describe the way one human being perceives what another says or does. States, or governments, may be made up of human beings. but the sum is an entirely different thing from the parts. Words invented to describe people and their behavior, including *trusting* and *trustworthy*, should not be applied to these other, nonhuman entities. To do so can be and often is grossly misleading.

In sum, no, we cannot trust the Soviet government, but not because it is made up of Communists and Russians. Our Founding Fathers long ago concluded that we individual citizens cannot simply "trust" our own government indefinitely either. That is why the Constitution they wrote restricts the powers of government and describes and reserves the rights of citizens.

Treaty writing and other state-to-state projects simply must take this—often unwelcome—fact into account.

MILITARIZATION
OF SPACE

Strategic Reconnaissance

WITH G. ALLEN GREB

T he early Eisenhower years saw the initiation of two different systems for conducting overhead reconnaissance. One was the U-2, a very high flying airplane. The other was the Agena satellite system, a rocket-powered platform designed to launch and support military payloads having a variety of purposes, including reconnaissance and surveillance.

Both of these programs were conducted behind an especially thick cloak of secrecy—key parts of them were conducted as so-called "black programs" not subject to normal budgeting and contracting procedures. They came to public attention only long after they were in regular use.*

Even so, the U-2 and the Agena satellite system have played very crucial roles in U.S. security affairs for the last two decades. Throughout this period they have provided most of the data on Soviet programs and deployments which has been used to generate our own plans. And more recently, satellite systems have provided the means for assuring compliance that is essential to all arms control negotiations and agreements.[1]

Aside from military applications, the military satellite programs have contributed the largest part of the industrial technological base underlying the U.S. space programs of the 1960s and 1970s, and have

---

* For a rare early discussion of the security aspects of the military reconnaissance program, see the testimony of L. Eugene Root of Lockheed Aircraft Corporation in the so-called Johnson hearings of 1957–58.

"In my opinion," Root advised, "if we do not take advantage of talking about this (military reconnaissance system), we could be directly accused of using it for overt purposes, and so we need to have a real vigorous but controlled program of releasing information." The advice was not followed. U.S. Congress, Senate, Preparedness Investigation Committee of Committee on Armed Services, Inquiry into Satellite and Missile Programs, 85th Congress, 1st and 2nd sessions (2 parts, 1958), part 2, p. 1855.

furnished a substantial majority of the program leadership as well.

The story of why and how these programs came to be provides a concise view in microcosm of the way in which government, industry, the universities, and non-profit institutions have interacted with each other since World War II to produce the major elements of the technological arms race. It is also a story that allows us to examine the extent to which critical national decisions and programs are the product of particular individuals—however brilliant, dynamic, persistent or persuasive—and to what extent they are the sum product of something which has simply grown naturally, almost inevitably, out of the political situation, the state of mind of the technologists, or the state of the art of technology.

The idea of aerial reconnaissance of enemy territory during wartime is as old as artificial flight itself. Yet the idea of conducting peacetime aerial reconnaissance to gather strategic intelligence was not seriously considered by the U.S. government until toward the end of the Truman administration, following the first Soviet atomic bomb test and during the Korean War. Then, in 1951–52, amidst growing fear of technological surprise and surprise attack, several closely connected and influential individuals, organizations, and ad hoc study groups took up the idea of aerial reconnaissance, expanded on it, and promoted it with higher governmental authorities.

One of these organizations was the Air Force Development Planning Office, then headed by General Bernard A. Schriever. Among his assistants was Richard Leghorn, an Air Force officer long interested in aerial reconnaissance. Leghorn is considered by many as the principal author of the modern (postwar) doctrines and ideas concerning tactical and strategic reconnaissance by means of overflight. His specific task in Schriever's office seems to have been to work out a plan for conducting such reconnaissance.[2] After leaving the Air Force in 1955, Leghorn became the nucleus of the group that formed Itek, a private firm that developed and built equipment of the kind used in such programs.

Among those whom Leghorn consulted, while still in the Air Force headquarters, was Amrom Katz. A civilian scientist working for the Air Force on aerial reconnaissance techniques since 1940, Katz moved to the RAND Corporation in 1954 where he participated in the work of that organization on reconnaissance and satellites. At the beginning of the second Nixon administration, Katz became a senior official in the Arms Control and Disarmament Agency, most of whose own action program depends on reconnaissance techniques.[3]

During this same period, in May 1951, the Air Force arranged for a special ad hoc study of the problem, known as Beacon Hill. This study, which took place in Boston, brought together for the first time in this connection several of the men who would for many years thereafter play key roles in these activities, including Edwin Land, founder and president of the Polaroid Corporation; James Baker, an astronomer and an exceptional designer of optical equipment; Edward Purcell, a Harvard physicist and Nobel laureate; and Allen Donovan of the Cornell Aeronautical Laboratories.[4]

Along with other sources, the Beacon Hill study made use of work done at a Boston University laboratory that had for some time been doing "state of the art" work on the optical equipment necessary for aerial reconnaissance. (Leghorn was also involved with this program.) According to Katz, the year-long study produced a "great report" on the subject of cameras for taking high altitude pictures.[5]

A third group that was active during that same critical time period was the Air Force Scientific Advisory Board's Panel on Physical Sciences. Established in 1951, this panel, whose original name was deliberately misleading, later became known as the Intelligence Systems Panel (1953–55) and the Reconnaissance Panel (1955–62). There was, as is very often the case, a considerable overlap in its membership with the other groups working in the area. Thus the panel's first chairman was George Kistiakowsky, the second was James Baker, the third was Carl Overhage, and Edwin Land and Allen Donovan were members from the beginning.[6]

Finally, as a result of all these and other studies, the Air Force took the first steps to set in motion a project to develop suitable cameras and the special aircraft needed to fly them.[7]

Four companies submitted proposals for developing very high flying aircraft appropriate for such a purpose. Two contracts were let, one to Bell Aircraft and one to the Glenn L. Martin Company, for relatively orthodox twin-engine planes. The intelligence panel of the Air Force Scientific Advisory Board (AFSAB) did not like those designs, however, and arranged for Donovan to visit a number of U.S. aircraft companies to learn what he could about alternative approaches.

One of the companies Donovan visited was Lockheed, and he was very favorably impressed by the particular proposal of Clarence L. "Kelly" Johnson for what became known as the U-2. Lockheed had previously submitted an unsolicited proposal for this same long-range reconnaissance airplane, essentially a single-engine jet-powered glider.[8]

An important new force intervened at this point which helped seal the decision to employ Johnson's design. This new force was the Technological Capabilities Panel (TCP), sometimes referred to as the "Surprise Attack Panel" or the first "Killian Panel." Organized in March 1954 by the Science Advisory Committee of the Office of Defense Mobilization

(SAC/ODM) at the direct request of President Eisenhower, the Technological Capabilities Panel was charged with an overall review of defense research and development programs. The director of the panel was James R. Killian, Jr., of MIT and the deputy director was James B. Fisk of the Bell Telephone Laboratories, both of whom were members of the parent Science Advisory Committee.

The actual detailed work of the Technological Capabilities Panel was done primarily by three subpanels. The first two dealt with strategic weapons and air defense and are not of direct concern for this discussion. Subpanel three, however, was chaired by the ubiquitous Edwin Land, and had as its main purpose a review of the need for and the means of gathering technological intelligence. Its report repeatedly stressed above all other considerations the need to apply the "creative resources of science, engineering, and technology" to the reconnaissance field.

"We must," the subpanel maintained, "find ways to increase the number of hard facts upon which our intelligence estimates are based, to provide better strategic warning to minimize surprise in the kind of attack, and to reduce the danger of gross overestimation of the threat."

Not surprisingly, subpanel three, which again included AFSAB panel members James Baker and Edward Purcell, who continued to be in close touch with Donovan, concluded that the Lockheed proposal—Kelly Johnson's U-2— was, in fact, the best. And they successfully urged that it, rather than one of the more conventional and more elaborate alternatives, be adopted. They were attracted by the notion that the U-2 was a "mosquito," manifestly less hostile than something bigger; and, contrary to Air Force thinking, they insisted that it be unarmed. Johnson himself made a good impression on them, and they backed the U-2 in part for this reason also.[9]

The U.S. Intelligence Board (USIB), then the highest coordinating board for U.S. intelligence activities, endorsed the project in November 1954. After a private meeting with Killian and Land, and with the USIB endorsement in hand, Eisenhower gave final approval. Because he desired that it be primarily a civilian rather than a military operation, Eisenhower assigned management and administrative responsibility to the CIA, a decision that put the program entirely outside the standard organizational mechanisms.

Even so, because of the expertise and facilities it possessed, the Air Force was inevitably involved, and Trevor Gardner, Assistant Secretary of the Air Force for R&D, became the responsible Air Force official.

The selection of Gardner helped to assure that the project would go forward with all possible speed. He was a young, energetic, intelligent and brash engineering executive who believed firmly that not enough was being done to adapt technology to the needs of American national security. In fact, one of the main stimuli behind the formation of the Technological Capabilities Panel in the first place was Gardner's concern that the United States was not doing nearly enough in this area. After entering the Pentagon in 1953, he had become deeply involved in advancing the missile development program and the U-2 presented him with yet another opportunity to put his ideas into practice.

In early December, CIA Director Allen Dulles informed Richard Bissell of his staff that the U-2 program would be his direct responsibility, and in mid-December Bissell and Gardner met together for the first time to work out the details of the program. As later recalled by TCP director Killian, "Bissell proved to be an extraordinary project engineer, and all of us who were associated with the undertaking...watched with great enthusiasm the skill which he brought to bear on making headway with the project."[10]

The project was conducted with a high degree of informality, and as a consequence the contractors for the project—Lockheed for the aircraft, Pratt and Whitney for the engines, and Hycon and Perkin-Elmer for the cameras—assumed an unusually central role in its planning and development. A small advisory group under Land's leadership continued to provide guidance to the project throughout its life, and has continued to play a similar important role in regard to all follow-on programs.[11]

The "Skunk Works," as Kelly Johnson's facilities were called, went to work immediately after the Bissell-Gardner meeting, and the first flight of a U-2 took place in August 1955, substantially less than a year after Eisenhower approved the program.

Whatever may be the legitimacy of these "black programs" in a democratic society, it is probable, as all of the insiders claimed, that the program could never have been accomplished in anything like this short a time if it had had to go through all of the formal budgetary and contracting procedures prescribed for open or just ordinarily secret programs.

Destabilizing Situation

The commonly shared perception of the growing menace posed by the Sino-Soviet bloc provided the motivation and justification for the proposal to construct and use the U-2. In so far as the specific proposal was concerned, the extremely closed nature both of the USSR and China was especially important. Indeed, the isolation of these two major countries from the rest of the world was unprecedented in modern times.

The pre-World War II situations in Nazi Germany and fascist Italy had been entirely different. True, some areas had been off limits, and some important activities—the V-2 program, for example—had been successfully hidden; but by and large those countries were open, and travel both to and from them was common. In marked contrast to that familiar situation, private communications with Russian in the last Stalin years and with China in the early Mao years were almost entirely cut off and, as we now know, the classic forms of espionage were producing only marginal results.

In modern strategic jargon such a situation is said to be highly "destabilizing." In such circumstances, it often become necessary to plan to cope with what the other side might be doing rather than what it actually is doing. And what it might be doing is limited only by human imagination rather than by physical reality.

American authorities were determined to somehow penetrate this situation, and it is only natural that when defense-oriented scientists and engineers faced this problem they emphasized the need for much more information about Soviet technology and focused on technological means for acquiring it.

No one involved seems to have seriously questioned the political morality of overflight. They thought of it in the context of spying, and in that context such clandestine programs were legitimized by custom. This attitude is evident, for example, in President Eisenhower's comments and reflections on the program:

> The U-2 reconnaissance program had been born of necessity. In the middle fifties the United States found itself, an open society, faced by a closed Communist empire which had lost none of its ambitions for world conquest, but which now possessed, in airplanes and guided missiles armed with nuclear weapons, an ever-growing capacity for launching surprise attacks against the United States. As long as the Communist empire remained closed, this capability would become ever more dangerous.
>
> ...The importance of the effort at that time cannot be overemphasized. Our relative position in intelligence, compared to that of the Soviets, could scarcely have been worse. The Soviets enjoyed practically unimpeded access to information of a kind in which we were almost wholly lacking...Considering all these things, I approved the recommendation of the intelligence chief that he employ the U-2...over Soviet territory.[12]

Even Leo Szilard, the U.S. nuclear physicist who often was among the first to see a moral issue in some new technological idea or event, was not seriously bothered by the fact of the overflights when they were later suddenly and unexpectedly revealed as a result of the successful Soviet interception of Francis Gary Powers' flight in 1960. Instead, he expressed "indignation, such as I have rarely experienced" only with respect to the way the U.S. government lied about the matter to the American people.[13]

Beginning in late June 1956, a series of six U-2 flights overflew European Russia, including Moscow and Leningrad. Following these first six flights, there was a brief pause resulting from a secret diplomatic protest by the Soviets, but they were resumed soon afterwards.

In this connection, it should be recalled that Eisenhower had proposed that the United States and the Soviet Union conduct limited but open overflights of each other's territory as a means of reducing the political tensions created by the growing fears of surprise attack. This idea, known as the Open Skies proposal, was put forth by President Eisenhower at the Geneva summit conference of July 1955 (fully six months after approval of the U-2 project), and at that time was summarily rejected by Khrushchev. [14]

A few months later, however, the Soviets did indicate they would be willing to consider certain carefully controlled and limited aerial reconnaissance of their territory as a "confidence building measure." Progress toward working out the differences between the U.S. and Soviet positions was, as usual, very slow and other events intervened before the differences could be resolved.

The U-2 Data

U-2 flights were launched from Turkey and Pakistan in order to photograph the missile development activities taking place at Kasputin Yar in south Russian, Tyuratam in Kazhakstan, and Sarishagan in central Asia, and to observe what was going on at the nuclear test range near Semipalatinsk. These and the earlier flights provided very useful information about the Soviet long-range bomber program, and they demonstrated that the U-2 had sufficient range and covered enough area to provide invaluable input about the development status of the Soviet missile program.

At the same time, however, they also showed that it was impossible for the aircraft to give adequate information about deployment of Russian missiles. Missile testing activities were conducted in a few well-known locations, but missile deployments could be almost anywhere within the 7 million square miles area of the Soviet Union.

If we arbitrarily suppose that the U-2 photography was effective over a band on the ground twice as wide as its altitude (14 miles), then a 3,000 mile long track would cover only about one percent of the area of the country, even before correcting for cloud cover, which would very substantially reduce its effectiveness. During the "missile gap" furor of the late 1950s, the U-2 was, in effect, being asked to prove a negative. And the coverage it could achieve was simply inadequate for that purpose.

On balance, therefore, the U-2 data alone was unable to allay growing concerns in Washington about the Soviet missile program. These concerns were much magnified when Sputnik suddenly confirmed to the public what the intelligence community already knew, that the Soviets had had a very substantial program of long-range missile development underway for some years and that it was just about to produce some important results.

The facts, and therefore the concerns also, were much exaggerated by some individuals within the scientific–technological community. They were able to exploit the mood of insecurity created by Sputnik, and the seemingly miraculous record of science since World War II, to gain unprecedented influence over American policy-makers, particularly over important segments of the Congress.[15]

The desire to obtain more and better intelligence information was at a peak, and hence it was this charged political climate which led the government to speed up the development and deployment of certain follow-on systems, most notably reconnaissance satellites, well before the downing of Francis Gary Powers' U-2 in 1960 signalled the end of that phase of the U-2 program.

Satellite Systems

From the time the notion of orbiting artificial or man-made satellites was first introduced in U.S. military circles at the end of World War II, the value of using such satellites as military observations posts was realized.[*] In fact, studies made by the RAND Corporation (initially a unit of Douglas Aircraft Company known as Project Rand) during the Truman administration not only demonstrated the feasibility of space reconnaissance but also offered fairly accurate estimates concerning the dimensions and other hardware.[16] No program for the development of such a system was initiated during the Truman years, however. This was mainly because the great cost of developing and building the rocket boosters needed to put the satellites in orbit simply could not be justified on the basis of this application alone.[17]

---

[*] The reconnaissance satellite system has had many different code names during its history, including WS 117L/Agena, Pied Piper, Weapon System 117L, Big Brother and Samos.

Early in the Eisenhower administration this situation changed radically. A combination of political events and some major advances in technology brought about the initiation of several programs for the development of very large rockets for use as long-range missiles. The political events were the end of the Korean War and the first complete change in personnel at the highest levels of the national security apparatus in a dozen years. These events resulted in a total review of U.S. national security policy, known as the "New Look," and that review in turn led to the proclamation of the Doctrine of Massive Retaliation and an increased emphasis on reliance on military technology rather than manpower.

The major technological advances were the invention of the hydrogen bomb by Edward Teller and Stanislaw Ulam, and the important though less dramatic improvements in rocket propulsion technology (mostly at North American's Rocketdyne Division) and guidance systems (mostly at MIT in Stark Draper's Instrumentation Laboratory). In addition, there was a growing awareness that the Soviets were, or soon would be, exploiting the same technological possibilities, and the notion that we were in a technological arms race with them began to take hold.

Out of this concatenation of events came a quick succession of authorizations for no less than five high priority programs to develop large long-range rockets: the Atlas in 1954; the Titan, Thor, and Jupiter in 1955; and the Polaris in 1956. The first of these authorizations came out of a review of the situation by a very special group, the Strategic Missile Evaluation Committee (SMEC, also sometimes known as the Teapot Committee). The principal figures in that review were Trevor Gardner, John von Neumann, George Kistiakowsky, Simon Ramo, and Jerome Wiesner. Each of the other authorizations flowed from a more complex set of circumstances which are not of direct interest to this discussion.

It was obvious at the time that the first four of the rockets listed above would be suitable for use as the main booster in a multistage rocket system capable of launching satellites of the size (1,000 pounds and up) described by RAND as being adequate for reconnaissance from space. And the first three rockets were, in fact, eventually used for such purposes.

On March 1, 1954, just a month after von Neumann's Strategic Missile Evaluation Committee (SMEC) had submitted its report to the Air Force urging the development of the first ICBM (Atlas), RAND issued a report on "Project FEED BACK," the then current name for the reconnaissance satellite proposal. This report, edited by James E. Lipp and Robert M. Salter, described all of the major subsystems in some detail, confirmed the validity and feasibility of the concept, and estimated that the development of a complete satellite system based on the existing state of the art would require seven years and cost $165 million.[18]

As a result of a combination of events—the publication of RAND's "Project FEED BACK" report, the initiation of a "highest priority" program to develop the Atlas missile; the growing attention being given to the importance of reconnaissance by such influential groups as the Air Staff, the Technological Capabilities Panel,[19] and the Air Force Science Advisory Board—the Air Force decided to go ahead with the program. On March 16, 1955, they issued "General Operational Requirement No. 80," a formal requirements document calling for the development of a reconnaissance satellite. Lockheed, RCA and Martin participated in the design competition that followed.

Lockheed's missiles and space division had already been interested in this project for some time. In order to strengthen its hand in the competition, Lockheed recruited a number of key people from RAND. The RAND people themselves saw the move as providing the opportunity to work on the actual design and construction of the systems they had been studying for some time, an opportunity which RAND itself by its nature could never provide.

Among those who joined Lockheed and who had long been involved in the RAND satellite studies were Louis Ridenour, L. Eugene Root (who shortly after became President of Lockheed Missiles and Space Company), Robert Salter, and James Lipp. Salter had been the author or a co-author of almost every major RAND report on the subject. Accordingly, he was able to formulate for Lockheed a proposal for a satellite system which in effect incorporated almost everything the Air Force knew and thought about the subject.[20]

Not surprisingly, then, the Air Force considered Lockheed's proposal the most satisfactory of those submitted and on Oct. 29, 1956 that firm was awarded a letter contract making it the prime contractor on what became known formally as Weapons System 117L. The Air Force Ballistic Missile Division (AFBMD) under General Schriever's command was given the responsibility for managing the project.

The Ramo-Wooldridge organization and the von Neumann committee, which were involved in all other AFBMD programs, were left out of the management loop in this case. At about the time the contract was awarded and "the money really started coming in," the analysts who had been working on the theoretical aspects of the program were shunted aside and a program management expert, John H. Carter, formerly a colonel on Schriever's staff at the BMD, became the program director.

In addition to a reconnaissance payload, the original 1956 contract also included the development of a system for conducting round-the-clock overhead surveillance of all large rocket launches by means of detection of the large amounts of infrared radiation emitted by such rockets during take-off. This particular subsystem became known as Midas and, after a somewhat fitful development program, it is today (1977) widely presumed to be in service as one of the major components of the American missile early warning system.

Agena Spacecraft

Since the Atlas alone could not put a significant payload in orbit,[21] a major component of the Lockheed program involved the development of a new additional rocket stage designed to sit atop the Atlas to provide the final push for getting the payload in orbit. This booster-satellite stage is known as the Agena. The first version was 19 feet long, 5 feet in diameter, and was powered by a Bell Aerosystems rocket engine capable of generating 15,000 pounds of thrust. Fully fueled, it weighed 8,500 pounds but its orbital weight (without propellants) was only about 1,700 pounds.[22]

The Agena A was the first large rocket whose engines were designed to be stopped and restarted in space, an absolute essential for providing any sort of capability to maneuver. This innovation, in turn, has allowed Lockheed to boast of myriad "firsts" achieved by various models of Agena spacecraft since 1959. Agenas "were first to achieve a circular orbit, to achieve a polar orbit, to be stabilized in all three axes in orbit, to be controlled in orbit by ground command, to return a man-made object from space, to propel themselves from one orbit to another, to propel spacecraft on successful Mars and Venus flyby missions, to achieve a rendezvous and docking by spacecraft in orbit, and to provide propulsion power in space for another spacecraft."[23]

On February 28, 1959, this Agena stage, sitting as a second stage on top of a Thor booster, launched the very first U.S. satellite having a weight (1,300 pounds) at all comparable to the early Soviet Sputniks. (Other U.S. systems prepared in conjunction with the International Geophysical Year and described below, had launched very small (31 pounds) payloads beginning Jan. 31, 1958.) The payload carried into orbit by that first successful Agena launch was known as Discoverer I.[24]

The stated purpose of Discoverer was to investigate various techniques necessary for the exploitation of space flight, including maneuvering in space and return to earth. From this point on, for a period of well over a year, the program ran into a singularly frustrating string of difficulties.

One after another, the satellites either failed to reach orbit, failed to behave correctly when they did, failed to respond to a command to return, or
were lost if they did. Only in August 1960 was Discoverer XIII successfully returned to earth and recovered near Hawaii by an Air Force task
group. It was followed in turn by many others in succeeding years. It has
been almost universally assumed that Discoverer's real purpose was pho-
toreconnaissance, but U.S. authorities have never officially confirmed that
assumption.[25]

The first attempt to launch the more powerful Atlas-Agena combination was made on Feb. 26, 1960, but it failed because of a malfunction at
the time the Agena stage separated from the Atlas. A second attempt on
May 24 was successful and put a 5,000 pound Midas satellite in orbit.[26]

Since these beginnings, the reconnaissance satellite program has pros-
pered very well. It is the main root of the confidence exhibited by American leaders (and the leaders of other countries as well) when they discuss
the numbers and other details that go into the nuclear "balance of terror"
equation. And in 1972, with the signing of the first SALT agreements, the
leaders of the rival blocs publicly endorsed the use of such satellite surveillance systems for monitoring international arms control agreements.

This endorsement is almost always put in euphemistic terms. For instance, the treaty limiting antiballistic missiles (ABMs) states simply that
monitoring of compliance shall be by such "national technical means of
verification" as are "consistent with generally recognized principles of international law." It is now universally accepted that overflight by satellite
is consistent with such principles, but overflight by aircraft is not.

The Agena stage and its descendants, and the Thor, Atlas and Titan
boosters, have continued to provide the means for launching practically
all U.S. military satellites and most large civilian satellites as well. The
only major exception is the Saturn V system used in connection with the
Apollo program; but even that huge rocket got the thrust required for its
first stage from engines whose development was originally started under
the aegis of the Air Force missile and space programs and only later trans-
ferred to NASA.

IGY Satellites

It is interesting to compare the timing of the secret military satellite program with the highly publicized International Geophysical Year (IGY)
program.[27] Because of the way information about these two programs was

handled—one highly secret, the other highly publicized—a false impression has been created about the relative timing of the two and their relative roles in laying the groundwork for the current U.S. space programs, both military and civilian. The details of the IGY program have been fully presented elsewhere.[28] In outline, the origins of the two IGY satellite systems—Project Orbiter and Project Vanguard—are as follows.

Project Orbiter, in which a Redstone rocket was to serve as the first stage booster of a system capable of putting a few pounds in orbit, was first seriously proposed by Navy Lt. Com. George Hoover and Army scientist Wernher Von Braun at a meeting in Washington on June 25, 1954. This was three months after the issuance of RAND's highly classified "Project FEED BACK" report but was done without knowledge of it. Somewhat later, another Navy group, centered in the Naval Research Laboratory, proposed a different booster system, based on the Viking rocket. Shortly after that, the U.S. National Committee for the International Geophysical Year urged that satellite flights be included as part of the U.S. contribution. And, finally in March 1955, Eisenhower approved the idea and assigned the responsibility for sorting out the proposals to the Pentagon.

A fair and competent review selected the Naval Research Laboratory's version, and Project Vanguard, as it came to be called, was duly authorized to proceed on Sept. 9, 1955. The promoters of Project Orbiter in the Army Ballistic Missile Agency fought back and, in effect, managed to keep their project alive under another guise, as the so-called "Jupiter-C" project, whose ostensible purpose was to check out ideas then circulating in the Army missile community about how to design reentry bodies. (It was, in fact, actually used for that purpose.)

On Sept. 20, 1956, the Army launched a Jupiter-C booster system which was complete in every way, except that—under orders from the highest levels in the Pentagon—the fourth stage was filled with sand instead of propellant. It was completely successful, and the odds are good that it would have gone into orbit on that date if the last stage had been appropriately charged.

On October 4, 1957, Sputnik went up, and soon after the Jupiter-C/Project Orbiter program was authorized to proceed in parallel with Project Vanguard. The first U.S. satellites to reach orbit were, as is well remembered, derived from the IGY program. But the first U.S. satellites having weights comparable to those of the first Soviet Sputnik came out of the military programs, and so did most of the personnel and the hardware used in the subsequent U.S. space programs.

The foregoing is written largely in terms of what certain leading technologists and public officials thought, recommended, and did. It is also instructive to consider the development programs of the U-2 and the reconnaissance satellites purely in terms of the status of the requisite technology. When viewed in this way, one finds that the developments under examination were approved at precisely the time the underlying technology matured and became available, and that the technology reached that state of readiness for reasons having nothing to do with reconnaissance objectives.

The example of the reconnaissance satellite is the clearest case in point. The development of the Atlas rocket was authorized in 1954 when U.S. intelligence indicated the Soviets were moving along similar lines, and when it became evident that the technologies for building the necessary thermonuclear warhead, propulsion system, guidance system, and space frame were all either in hand or clearly in sight.

When it became more widely evident a year later that such a rocket could and would really be built, the Air Force then issued its requirement for a reconnaissance satellite of the type and weight that could be launched into orbit using such a first stage booster plus a special upper stage based on very similar technology. As pointed out above, the idea of a reconnaissance satellite had been present for nearly a decade, but a program to build one had not been authorized previously because the every expensive and difficult booster development could not be justified for that application alone.

The U-2 presents a similar though less obvious case. Airborne reconnaissance equipment had been under development for a long time, but no aircraft that was capable of flying high enough and far enough and that was not at the same time very threatening existed before the U-2. And the U-2 airplane became possible only after a suitable jet engine, Pratt and Whitney's J57, had been developed for other reasons.[29] Whether a similar aircraft would ever have been built without the overflight requirement is uncertain, but the converse is fairly certain: it could not have been built much sooner than it actually was no matter how firm a requirement for long-range overflight there might have been.

In summary, then, both the U-2 and the reconnaissance satellites appeared precisely when the general forward movement of technology made it possible for them to do so. Does this in turn mean that all the persons named above were, as individuals, unimportant? Were they just puppets in a world totally governed by technological imperatives?

We think the most fruitful way to look at the matter is this: the general Cold War situation plus the high level of technological exuberance that has characterized American defense thinking since World War II created implicit niches for these people. The individuals who filled the niches had both the opportunity and the intellectual ability and energy to do so.

This situation is not different from most success stories. The opportunity that make success possible comes along in large part by chance, but making something of the opportunity does depend on the personal characteristics of the person involved. Each of the individuals in this story influenced the details of the events that took place according to the dictates of his own individual technological and political personality; but each of them could have been replaced by any one of a small number of not too dissimilar individuals. The overall result would very likely have been the same.

What was important, then, is not that these particular men were active in American technology at that particular time, but that there was a pool of such people available—perhaps just barely large enough, perhaps adequate several times over. Without such a pool, a technological imperative could not per se have produced the results described.

It was not only the state of the art of American technology that mattered; the state of mind was equally important as well. And that state of mind was the collective creation of the principal actors and their colleagues.

In the same vein, how is one to assess the role of President Eisenhower? Neither the U-2 nor the reconnaissance satellite project originated with the President or his immediate political advisers. Rather, they were urged on him by technical advisers who were convinced both of their efficacy and feasibility. Given the Cold War situation, and particularly the growing fear of a possibly decisive surprise attack, it is difficult to believe that any man who might have been elected President at that time could have turned down the proposal to build a reconnaissance satellite. In this case, the role of the President as decision-maker seems to have been less than the role of his chief technological advisers.

But the case of the U-2 is more complicated. It required a special mixture of boldness and self-assurance on the part of the President. Nearly all Americans in the 1950s implicitly but firmly believed that U.S. goals were moral, that Soviet goals were evil, and that this difference was obvious even to the Soviet leadership itself. Again, any president would have shared in this national mood. But, because of his wartime experience and reputation, Eisenhower in particular demonstrated an unusual degree of certainty and confidence when confronted with problems of a military nature such as the U-2 overflight program entailed.

In this case, then, it does seem as if the President's personal leadership (and his personal concern about a possible surprise attack), the state of the art, and the state of mind of American technology were all separately crucial in initiating this particular project.

References

1. For an excellent discussion of reconnaissance and its relation to arms control, see Ted Greenwood, "Reconnaissance, Surveillance and Arms Control," Adelphi Papers, no. 88 (London: International Institute for Strategic Studies, 1972).

2. Among other military figures prominent in promoting the idea of reconnaissance and developing important aerial techniques in the post-World War II years were Colonel R. W. Philbrick, Colonel Karl Polifka, and Brigadier General George W. Goddard. See Amrom H. Katz, "A Tribute to George W. Goddard," Airpower Historian, 10 (Oct. 1963), 101–09; and A.H. Katz to York and Greb, July 1, 1976.

3. Katz, Some Notes on the History of Aerial Reconnaissance (Santa Monica, Ca.: RAND, April 1966); "Observation Satellites: Problems and Prospects," Astronautics, 5 (April, June–Sept. 1960); and "Thoughts on Reconnaissance," in Selected Readings in Aerial Reconnaissance, edited by Katz, report P-2762 (Santa Monica, Ca.: RAND, Aug. 1963).

4. Others included Carl F.J. Overhage, chairman of the Beacon Hill group and later director of the Lincoln Laboratory, and Gilbert W. King, later a vice president of the Aerospace Corporation. Among the part-time consultants to the study were Jerome B. Wiesner, now president of MIT, who became Special Assistant for Science and Technology under President Kennedy, and Louis N. Ridenour, one of the authors of the original 1946 RAND study on satellites.

5. Amrom Katz, personal interview, 1975.

6. Other members included Milton Plesset of RAND and Herbert Scoville, later Deputy Director of the CIA and Assistant Director of the Arms Control and Disarmament Agency.

7. According to Eugene Emme, NASA historian, at least one project involving reconnaissance overflights of the Soviet bloc predated the U-2 promotion. This project apparently had the code name "Moby Dick" and utilized very high altitude balloon-borne cameras but it was never effective.

8. L. Eugene Root, then with Lockheed and before that in RAND's Washington office, was almost certainly aware of the Air Force ideas about the need for very long-range reconnaissance aircraft. Allen Donovan, personal interview, Sept. 1976.

9. The Technological Capabilities Panel report itself, issued in February 1955, has only recently been declassified except for those parts dealing with intelligence. See The Report to the President by the Technological Capabilities Panel of the Science Advisory Committee (2 vols.; Feb. 14, 1955), Gordon Gray Papers, Eisenhower Library; the quotes are from vol. 1, pp. vi, 24–25, and 44.

10. James R. Killian, Jr., to York and Greb, Oct. 25, 1976.

11. The case of the U-2, it should be added, is an example (an extreme one to be sure) of an organizational style characteristic of the Eisenhower administration. It seems that the research and development phase of almost every major strategic system approved by the President—for example, the Atlas-Titan and Polaris systems—was removed from the conventional service structure.

12. Dwight D. Eisenhower, The White House Years: Waging Peace, 1956–1961 (Garden City: Doubleday, 1965), 544–45.

13. Leo Szilard, letter to the editor, New York Herald-Tribune, May 10, 1960. In a second letter published June 5, Szilard also asked whether we "[w]ould not be better off if the Russians could regard the[ir] rocket bases as 'invulnerable,' by virtue of our ignorance of their location? Would not our ignorance of the location of these bases eliminate such rational arguments, as might otherwise be cited, in favor of Russia staging a surprise attack against our rocket, and strategic bomber, bases?"

14. "United States Outline Plan for the Implementation of President Eisenhower's Aerial Inspection Proposal," in U.S. Department of State, Documents on Disarmament, 1945–1959 (2 vols.; Washington, D.C., 1960), I, 501–03; Bernard G. Bechhoefer, Postwar Negotiations for Arms Control (Washington, D.C.: Brookings Institution, 1961), 301–76 passim.

15. Cf., Eisenhower, Waging Peace, p. 547.

16. The first of these RAND reports appeared in 1946 under the title Preliminary Design for an Experimental World-Circling Spaceship, report no. SM-11827 (Santa Monica, Ca.: Project RAND, May 2, 1946). For a discussion of this and other reports, see Philip J. Klass, Secret Sentries in

Space (New York: Random House, 1971) chap. 8; R. Cargill Hall, "Early U.S. Satellite Propos-
als," Technology and Culture, 4 (Fall 1963), 410–34. The latter also in The History of Rocket
Technology: Essays on Research, Development, and Utility, edited by Eugene M. Emme (Detroit:
Wayne State University Press, 1964), pp. 75–79.

Klass was editor of Aviation Week all during the period involved. He had no clearance, but he
had excellent access to various mixtures of rumor and fact, especially to those facts which were
self-serving for their sources. As a result, while the book does contain some factual errors, it is basically and generally quite reliable.

17. A study made in 1952–53 at the direct request of President Truman mentioned possible military,
as well as scientific and psychological, applications of satellites but recommended only that "a
small but effective committee" of engineers, scientists, and State and Defense Department officials be created to study the problem. Aristid V. Grosse, Report on the Present Status of the Satellite Problem (Philadelphia: Research Institute of Temple University, Aug. 25, 1953), pp. 6–7,
MDHC files, Truman Papers, Truman Library.

18. The report, officially titled Project FEED BACK, Summary Report (2 vols., March 1, 1954) still
carries a "Confidential" classification. Some 200 people are said to have been involved in its
preparation. They included Bruno Augenstein, who had played a major role in generating
RAND's ICBM studies, which in turn had had considerable influence both on SMEC and on the
Air Force's decision to go ahead with larger rockets. Augenstein later (1964) became a Special
Assistant for Intelligence and Reconnaissance in the office of the Director of Defense Research
and Engineering.

19. As the authors understand it, the TCP subpanel did not like certain specific engineering details of
the RAND proposal, and apparently did not endorse it per se; but the Technological Capabilities
Panel's general endorsement of overhead reconnaissance was a very positive factor in promoting
its acceptance.

20. As one former Lockheed official put it, "Salter's central role since the beginning enabled him to
write a proposal in a form which allowed the Air Force to tear off the Lockheed covers, substitute
their own covers, and label it the ARDC (Air Research and Development Command) Development Plan." Private communication, Oct. 2, 1973.

21. That is, the large-scale, super-precise optics required for obtaining suitable pictures from approximately 100 miles overhead. Such equipment could not be usefully packaged in a satellite with a
weight under 1,000 pounds.

22. John W. R. Taylor, ed., Jane's All the World's Aircraft, 1964–65 (New York: McGraw-Hill,
1964), p. 426. Klass (Secret Sentries, pp. 83, 92) gives slightly different figures for weights:
launching weight, 8,000 pounds; weight in orbit, 1,300 pounds.

23. John W. R. Taylor, ed., Jane's All the World's Aircraft, 1975–76 (New York: McGraw-Hill,
1975), p. 662.

24. In December 1958 an entire Atlas was placed in orbit. This isolated event, known as "Project
Score," was carried out strictly for propaganda purposes. The gross weight was indeed large
(8,500 pounds), but its so-called useful payload was a very small communications-relay package
hastily arranged for the purpose. It was never followed by any descendants, and it is more properly thought of as a rather peculiar Atlas test than as a part of the U.S. satellite program.
It should also be noted that there is some doubt as to whether Discoverer I actually achieved orbit.
Perhaps because of a total communications failure, no data was ever received from it. If Discoverer I did not in fact reach orbit, then Discoverer II, launched six weeks later on April 13, 1959,
was the first heavy U.S. satellite.

25. According to Klass (Secret Sentries, p. 101), the overall direction of the Discoverer system also
was shared by the Air Force and the CIA. The rationale behind this arrangement was the same as
in the case of the U-2, and the same small group of advisors played an important role in working
out the details of this program as well.

26. Dates, weights, names from Eugene M. Emme, Aeronautics and Astronautics: An American
Chronology of Science and Technology in the Exploration of Space, 1915–1960 (Washington,
D.C.: NASA, 1961).

27. The International Geophysical Year is an 18-month (July 1957 to Dec. 1958) cooperative effort

among all the major nations of the world for the purpose of studying the Earth and its atmosphere in a coordinated fashion. The U.S. National Committee was organized and sponsored by the U.S. National Academy of Science and is funded by the National Science Foundation.

28. One of the best accounts is R. Cargill Hall, "Origins and Development of the Vanguard and Explorer Satellite Programs," Airpower Historian, 11 (Oct. 1964), 101–12. Much of the following discussion on IGY satellites is based on this article. See also U.S. Congress, Senate, Preparedness Investigating Subcommittee of the Committee on Armed Services, Inquiry into Satellite and Missile Programs, 85th Cong., 1st and 2nd Sess. (2 parts, 1958); Kurt K. Stehling, Project Vanguard (Garden City: Doubleday, 1961); Constance M. Green and Milton Lomask, Vanguard: A History (Washington, D.C.: NASA Historical Series, 1970); John P. Hagen, "The Viking and the Vanguard," Technology and Culture, 4 (Fall 1963), 435–51.

29. The first production J57 became available in February 1953. The U-2 used a later version of this engine, the J57C with various modifications, including compressor blades to make it more suitable for operation at very high altitudes. John W. R. Taylor, ed., Jane's All the World's Aircraft, 1962–63 (New York: McGraw-Hill, 1962), pp. 235, 491–92.

Nuclear Deterrence and the Military Uses of Space

On March 23, 1983 President Reagan made a speech that was almost as unexpected as it was portentous. In it, he called on American science to create a total defense against ballistic missiles, a defense of such qualitative difference that it could support a radical change in our nuclear strategy. Nuclear strategy would no longer be based on offense, but rather on defense; no longer on assured destruction, but on assured survival; or, as some have put it, no longer on death, but on life.

The official name of the program is the Strategic Defense Initiative (SDI), but because the technical ideas underlying it involve deploying defensive weapons in space, the program has come to be referred to—even by most of its supporters—as "Star Wars."

Although the *strategy* involved in the SDI represents a radical departure from past thinking, the *technology* that underlies it is the direct and logical result of many decades of research and development. Indeed, proposals for using satellites to supplement earth-based military capabilities were first put forth seriously in the mid-1940s. Ten years later, both the United States and the Soviet Union initiated programs designed to implement certain of these ideas. For the next quarter of a century, the development and deployment of these military space assets, as they came to be called, proceeded in a roughly balanced and relatively non-threatening manner. Recent developments—including anti-satellite weapons, the direct use of space assets in warfare, as well as the president's Strategic Defense Initiative—are bringing about a new, more costly, and more dangerous phase in this ongoing process.

The Evolving Military Uses of Space

As early as 1946, a group of scientists and engineers at the RAND Corporation published a report entitled "Preliminary Design of an Experimental World-Circling Spaceship." The report was remarkable for its accuracy and prescience. Most of it is devoted to the question of how to build adequate launching systems, and while its estimates in this regard are not perfect, they are very good. More important, the report also devoted some attention to the possible military applications of satellites:

> ...The military importance of establishing vehicles in satellite orbits arises largely from the circumstances that defenses against airborne attack are rapidly improving....Under these circumstances, a considerable premium is put on high missile velocity, to increase the difficulty of interception.

> This being so, we can assume that an air offensive of the future will be carried out largely or altogether by high-speed pilotless missiles...[accordingly] precise observations of the position of the missile can be made from the satellite, and a final control impulse applied to bring the missile down on its intended target. This scheme, while it involves considerable complexity in instrumentation, seems entirely feasible.

> Alternatively, the satellite itself can be considered as the missile. After observation of its trajectory, a control impulse can be applied in such a direction, and at such a time that the satellite is brought down on its target.[1]

We have here the germ of two important ideas that were realized only decades later. The first finds its modern expression in the current plan to use navigation satellites of the global positioning system (GPS) to provide prompt location and steering information to a variety of weapons systems, including aircraft and missiles, thus increasing their accuracy and lethality. The second is the notion of an orbiting weapons platform, an idea implemented by the Soviets in their fractional orbital bombardment system (FOBS) in the early 1960s.

Other proposed military uses for satellites advanced in the 1946 report included locating targets, observing weather conditions over enemy territory, and using satellites to relay military communications. The report specified that satellites in geosynchronous orbits—36,000 kilometers above the earth—would be best for this last purpose.

Since no rockets suitable for launching a satellite into orbit existed at that time, it was soon concluded that the potential benefits of satellites would not by themselves justify the very large costs of the development and construction of the necessary rockets. For the next several years, then, work on satellites proceeded on a research basis only. As a result of a series of unexpected events, however—including, especially, the first Soviet A-bomb test, the Korean War, and the discovery of the existence of a substantial Soviet program dedicated to the development of long-range rockets—the U.S. in 1954 decided to establish several parallel programs aimed at developing huge rockets (Atlas, Titan, Thor, and Jupiter) specifically designed to power large and very long-range ballistic missiles. These rockets, of course, proved to be just right for the launching of artificial satellites, and as a result the idea of building such devices took on new meaning. By this time, satellite designers, still concentrated primarily at RAND, had begun to focus on those goals—overhead reconnaissance and early warning—that they felt were most promising for near-term practical use. In short order, a RAND study entitled "Project Feedback" was published on March 1, 1954, delineating these possibilities, followed a year later by a U.S. Air Force contract calling for the development of reconnaissance satellites.

All of these developments were, of course, proceeding on a highly classified basis. At about the same time, but in the public sphere, plans were being elaborated to declare July 1957–58 an "International Geophysical Year." One of the highlights of the year was to be the launching of artificial satellites that would perform scientific observations of our planet. In 1954 and 1955, work began on two additional satellites: a much smaller scientific satellite called "Project Orbiter," and the Navy's "Project Vanguard" (an alternative to "Orbiter"). However, before either of these programs could place a satellite in orbit, the Soviets, to the great shock and surprise of the general public and the Congress, launched the first Sputnik on October 4, 1957.

The launch of Sputnik prompted the U.S. government to undertake a thorough, and often frantic, review of the state of "high-technology" in America. Out of this came a total revamping and expansion of the then much smaller, but very prominent, civilian space program and the creation of a new agency, NASA, to oversee it. The still very secret military satellite program, on the other hand, experienced only minor changes in either its substance or organization. Like many other "high-tech" programs of the time, however, its priority and funding were eventually substantially increased.

This brief review of the first steps towards the military use of space exposes a fact that is commonly missed or glossed over in the folk-historic views of the origins and nature of our space programs: our space programs, from the beginning, have been primarily of a military, not a civilian or scientific nature. Public attention may have focused on Mercury, Gemini, and Apollo and the corresponding Soviet projects, but the fact remains that both the U.S. and Soviet space programs overall have always been driven more by military considerations and requirements than civilian and scientific ones.

Moreover, the military space program has always included a component dedicated to warfighting, as well as components devoted to general intelligence, administration, and logistical support. Our strategic missiles, for example, are currently primed to attack some 10,000 military and industrial targets, mostly in the Soviet Union. For all practical purposes, every one of these targets was discovered and located by reconnaissance satellites and was given precise coordinates by means of data produced by geodetic satellites. Our missile submarines derive their initial coordinates from data provided by navigational satellites. In the future, the use of these global positioning system (GPS) satellites will markedly improve the effectiveness of aircraft and missiles by correcting their trajectories while in flight. Mapping satellites are currently providing the input data for the super-accurate "tercom" (terrain comparison) system that will steer our cruise missiles. And we are already in the early stages of supplying data gleaned by satellites directly to battlefield commanders.

Ambiguity of Military Space Missions

Setting aside for the moment consideration of ASATs and the SDI, are the various elements of today's military space program stabilizing or destabilizing? Are they benign or aggressive? Are they intended to reinforce deterrence, or is their primary purpose to improve our fighting capability should deterrence fail? The answers to these questions are by no means simple or straightforward. Let us first consider overhead reconnaissance in the broadest sense. Reconnaissance is used for several major purposes, each of which affects the stability of the relationship between the superpowers. One purpose is target discovery and location—the "spotting" function of the 1946 RAND report. Another is general intelligence to evaluate the state of preparedness of an enemy. A third is verification of arms-control agreements. This last purpose receives most of the attention in those political circles primarily interested in arms control, detente, and diplomatic approaches to solving the world's problems. But it is the first two purposes that provide the main rationale for spending billions of dollars annually to operate the existing system and to develop new ones. General intelligence-gathering and the verification of treaty compliance are, by and large, stabilizing. Target "spotting" is not. On the one hand, the discovery by satellites in 1961 that there was in fact no "missile gap" between the U.S. and the Soviet Union prevented us from over-reacting to Sputnik even more than we did. In this case, satellite intelligence contributed to stability. On the other hand, these same satellites are also capable of providing data that is sufficiently complete and accurate to make a successful preemptive first strike against land-based military forces seem possible.

Early-warning satellites are similarly ambiguous in their function. They are most important for their ability to give reliable warning of a missile attack twice as quickly as conventional earth-based radars, thus making it more difficult for the enemy to achieve a successful first strike. For this reason, early warning satellites are said to be stabilizing. At the same time, these same characteristics are important to the launching of a retaliatory strike before absorbing a Soviet attack (launch-on-warning), a tactic advanced by some strategists as the best and cheapest answer to a surprise first strike on our land-based force. By making early warning more accurate, these satellites make launch-on-warning more thinkable.

The data produced by earth-mapping satellites give modern cruise missiles their exceptional accuracy. Air-launched cruise missiles have as their primary purpose extending the life of the bomber leg of the strategic triad; they are therefore commonly said to contribute to strategic stability because they reinforce deterrence in a fundamental way. Other types of cruise missiles, such as the ground-launched cruise missiles deployed in Europe, add very little to deterrence itself; they are primarily warfighting devices designed to destroy specific military targets in the event deterrence fails.

Similarly, the GPS, the Soviet radar ocean reconnaissance satellite (RORSAT), and other systems designed to supply critical information rapidly to tactical commanders, are clearly much more sharply oriented towards warfare than towards deterrence. Indeed, these systems belong to a special class of space weaponry that one American official has described as the "most important class of weapons in space," yet is "almost totally neglected by the arms-control community."[2]

Space, then, long militarized and weaponized, is becoming steadily more so. Up to now, however, these developments have taken place in a roughly balanced fashion, in which the actions of one superpower have not generally been construed by the other as immediately menacing or even particularly hostile. The reason for this is not that satellites have made no direct contribution to warfighting, but simply that they have so far done so only by a slow accumulation of data during peacetime. This situation is obviously about to change: new systems—GPS, RORSAT, and the direct use of space-based assets by tactical commanders—will soon be supplying essential data to warfighting systems in real time. Such a move will inevitably be seen as hostile and dangerous by both sides; a clear example of this is the U.S. Navy's strong reaction to the Soviet RORSAT, whose principal purpose is to locate and target U.S. surface ships.

In what way, then, would the deployment of a serious anti-satellite (ASAT) capability fit into this existing situation?

ASATs and Military Satellites

The pressure to acquire an anti-satellite capability is a natural response to the superpowers' increased use of space-based assets for military purposes. Indeed, the very first anti-satellite system was deployed by the United States during the Kennedy administration as a direct and immediate response to two seemingly ominous events in the Soviet Union. One of these was the development of an orbital bombardment system, that is, a system in which bombs would be deployed on satellites in low orbits and either exploded in space or brought down and detonated at specific points on the ground. The other was the explosion of an extraordinarily large nuclear device in 1961 at Novaya Zemlya in central Asia. While its estimated size was "only" fifty-eight megatons, American scientists believed the Soviets could have easily increased the yield to one hundred megatons or more. Many analysts concluded that bombs of this type could, if exploded at an altitude of one hundred miles or so, set fires on the ground over an area of perhaps a hundred thousand square miles—the area of a typical midwestern American state.

As a natural response to the threat posed by the combination of these two developments, Secretary of Defense McNamara ordered the deployment on Johnston Island of an ASAT system that used a Thor rocket to deliver a nuclear warhead to an altitude sufficient for intercepting bombs in orbit. While the Soviets did in fact eventually deploy a so-called fractional orbital bombardment system (FOBS), and although both the Soviet FOBS and the U.S. anti-satellite system remained deployed for more than a decade, each was eventually decommissioned.

Despite this instance of provocation and response, U.S. authorities during this period declined to authorize the development and deployment of a general purpose anti-satellite system. The rationale was simple: the U.S. relied heavily on satellites for intelligence and support purposes, and it was in our direct interest to maintain the idea of space as a special sanctuary for such systems. If we wanted no Soviet ASAT, we were willing to pay the necessary price: no American ASAT.

The Soviets, however, did not practice similar restraint. They initiated testing of a general-purpose ASAT in 1968.[3] After almost a decade of further restraint on our part, the Carter administration in 1977 finally responded to this Soviet ASAT project with a three-pronged approach. The first was the initiation of an ASAT development program of our own. The second was an increased effort to find ways of defending satellites against attack, the so-called DSAT program. Last was the initiation of negotiations with the Soviets designed to eliminate ASATs altogether, if possible, and hence to render the first two responses unnecessary. The Soviet reply was carefully ambiguous, indicating a preference for a "rules of the road" approach (i.e., regulating satellite use) over elimination of ASATs. Indeed, the Soviets made it clear that they regarded ASATs as a natural response to threats from space, mentioning in this context "third countries" (i.e., China) and "threats against our sovereignty from space." As it now stands, development of a general-purpose ASAT is going forward on both sides, despite the resumption in March 1985 of U.S.–Soviet negotiations on a broad range of weapons.

The main argument of those who favor ASAT arms control centers on the special uncertainties and dangers that the introduction of general-purpose ASATs will give rise to. So far, warfare, at least in the direct sense, has not extended to space; indeed, we have treaties now in force that were specifically designed to maintain outer space as a zone of peace. The introduction of ASATs, its critics contend, would destroy all that. Satellites whose purposes are largely benign and stabilizing—that is, those we depend on for intelligence, warning, communications, and the like—would be placed in peril and subject to destruction at any time. These critics foresee not only needlessly heightened fears and tensions, but increasing expenditures of money that could otherwise go to more useful military programs or neglected civil needs.

ASAT supporters argue that space will soon harbor even more satellites that contribute directly to an enemy's warfighting capability, and that it makes no sense, at this late date, to confer on space special sanctuary status. American officials cite the existence of Soviet ocean reconnaissance satellites (RORSAT) and contend that other systems, presumably parallel to those that we are developing, are in prospect.[4] They point to the difficulties of verifying any ban on ASAT deployment, and argue that new non-nuclear ABM systems (such as the one tested by the U.S. Army in 1984 at Kwajalein Atoll), to say nothing of powerful ground-based lasers and the like, make a treaty prohibiting ASATs too late and impractical.

Given these diametrically opposed positions, how is the American body politic likely to react? One basic fact to consider is that only one-third of the Senate is necessary to block any treaty, and no arms-control treaty that is openly opposed by the armed forces can make it over this hurdle. Although our military leadership is currently split on this issue, a growing majority appears to belong to the camp that regards ASATs as a natural response to a coming threat. Given that the most influential members of the office of the secretary of defense have similar sentiments, it is unlikely that a treaty totally banning ASATs will be negotiated in the near future.

On the other hand, a treaty establishing clear-cut "rules of the road" could probably be negotiated, even in the present circumstances. The world has had such rules governing behavior at sea for centuries. The oceans are full of fighting machines, yet these co-exist easily with civilian ships during peacetime. A similar arrangement could surely be worked out for space, one that would give all parties the same kind of assurance of safe passage for military as well as civilian satellites during peacetime that is now the case for all nations' warships now at sea. Such a policy could go a long way in guaranteeing the survival of intelligence and warning systems, as well as other systems that add to both international stability and military readiness in peacetime (in the event of a major war, of course, all space assets would be in grave danger). As with all high-technology weapons systems, especially nuclear systems, it is those components that play a major role in deterring war that should command primary attention, not those that would be important in combat, should deterrence fail. Of course, a total ban on ASATs would be the best way to reassure the continuing role of our space assets in deterring war, but a "rules of the road" agreement, if carefully crafted, could accomplish this almost as well, and from a practical political perspective seems much more feasible.

Space and Strategic Defense

The president's Strategic Defense Initiative was introduced into a world in which the militarization of space by the United States and the Soviet Union was already solidly underway, and in which it was widely recognized that this process ought to be limited or contained before it escalated beyond our control. Indeed, the technological side of the SDI is based, to a large extent, on concepts already deployed or currently under development in military space programs. Approximately 90 percent of the 1985 SDI budget is devoted to continuing the development of devices and the exploration of ideas that were already programmed before the SDI was introduced in March 1983.

For all practical purposes, the technological side of the SDI consists of those ideas, devices, and systems discussed in the report of the Fletcher panel, one of three committees set up by President Reagan to investigate strategic defense.[5] This report called for the creation of a multilayered defense system made up of a variety of components designed to detect, target, and destroy attacking ballistic missiles and their nuclear warheads all along their trajectories from their initial launch to their final targets. As noted above, most of the systems involved were based on concepts either already employed in our military space program or in various stages of development for possible future space applications. Many others were taken from terminal ballistic missile R&D programs established over a quarter of a century ago. And some came from unrelated programs, particularly from the research laboratories of our defense establishment. Among these are the so-called "Third Generation Weapons," including laser and particle beams and kinetic energy weapons that are considered especially useful in a defense mode against Soviet missiles in boost phase, immediately after they have been launched.

There is, or course, much debate over whether the SDI will result in a total defensive system possessing the characteristics and qualities necessary to bring about the strategic revolution the president called for. We can here summarize this debate by examining two quotations from expert studies of the issue. On the one side, the Fletcher panel noted that: "By taking an optimistic view of newly emerging technologies, we concluded that a robust BMD system can be made to work eventually. The ultimate effectiveness, complexity, and degree of technical risk in this system will depend not only on the technology itself, but also on the extent to which the Soviet Union either agrees to mutual defense arrangements and offense limitations, or embarks on new strategic directions in response to our initiative."[6]

On the other side, in a report written by Drell, Farley, and Holloway, we find the statement "We do not now know how to build a strategic defense of our Society that can render nuclear weapons impotent and obsolete as called for by President Reagan. Nor can one foresee the ability to achieve the President's goal against an unconstrained, responsive threat."[7] Many other studies by knowledgeable experts have reached very similar conclusions.[8]

Note that experts on both sides agree that the present technology cannot produce a system adequate to the president's goals; the disagreement is over whether the technology that will become available in the next ten or twenty years will be adequate to provide a truly useful level of defense, and whether the best way to generate such technology is to proceed along the lines now being proposed in the Department of Defense. The argument is actually over the future course of technology: to what degree we can extrapolate from existing know-how to estimate our capabilities in the distant future.

Space Weapons and Nuclear Deterrence

During a press interview conducted soon after his "Star Wars" speech, President Reagan was asked about his new initiative. At one point he replied:

> To look down on an endless future with both of us sitting here with these horrible missiles aimed at each other and the only thing preventing a holocaust is just so long as no one pulls the trigger...this is unthinkable.[9]

In giving this answer, the president, perhaps unwittingly, joined a host of others who believe that peace through a balance of terror is untenable in the long run, that deterrence through the threat of mutual assured destruction (MAD) cannot be the basis for an enduring peace, that the threat to kill hundreds of millions of civilians as punishment for some unacceptable political act by their government is immoral in the deepest sense of the word. These ideas are not new; all previous U.S. presidents have held similar ideas, and as far as we can tell, so have all Soviet leaders, at least since Stalin. The problem is—and has been for forty years—that no viable alternative presents itself. Supporters of nuclear deterrence have often pointed to the fact that, whatever the long-term prospects may be, deterrence has so far "worked"; that peace in Europe has been preserved for more than forty years, a time during which any number of crises might otherwise have precipitated a major war.

The introduction of nuclear explosives and other high-technology weapons only a few decades ago resulted in a radical change in the vulnerability of both the United States and the Soviet Union. Before that, the oceans that had from our origins isolated us from the rest of the world still protected us from both invasion and attack. And in 1941, when the Nazi war machine attacked the Soviet Union, the great size of that country enabled it to absorb the full weight of attack long enough to rally, muster outside help, and turn the tide of the invasion. Today, each of the superpowers could, if it chose to, literally annihilate the other in less than an hour, and the only recourse either nation has to forestall such a horror is to threaten to wreak revenge in kind. That is the stark meaning of Reagan's obviously heartfelt remark that the only thing preventing a nuclear holocaust is the fact that no one is pulling the trigger. Nor is this stalemate a static situation; things are in fact becoming steadily worse. The thoroughness of the potential destruction is increasing, the chance of survival for individuals and for civilization itself is decreasing. The combination of technical improvements in accuracy and multiple warheads (MIRVs) is making the possibility of a first strike seem, at least to some, more thinkable. To guard against this, some strategists advocate a policy of launch-on-warning, though such a strategy carries with it the possibility of accidental war. And last but not least, the ability to wreak this sort of total devastation is slowly but steadily spreading to more countries.

In sum, for forty years now our military power, as measured by our ability to wreak death and destruction on an enemy in an even shorter time, has been steadily increasing. At the same time, our national security, as measured by the combination of the ability of others to wreak death and destruction on us and our inability to do anything about it, has been steadily decreasing. This has been, and remains, our mutual and fundamental security dilemma.

Spiritual leaders such as the Pope and the American and French Catholic bishops have perceived the issue not very differently from our political leaders. While they have all condemned the use of the threat of nuclear annihilation as the principal means of maintaining peace, they have in various ways also said that this can be tolerated so long as we have no alternative way of avoiding large-scale war. More important, they have added that the current situation is tolerable only so long as we are actively and seriously seeking alternatives. Not everyone may agree with the way these spiritual leaders have stated their case, but nearly everyone believes that we should indeed be seriously seeking alternative ways of maintaining stability and assuring peace.

Our lives, the lives of our progeny, and the continuation of our civilization do depend on "nobody pulling the trigger." That, truly, "is unthinkable," and necessarily evokes extreme proposals for a solution. Some people suggest panaceas: general and complete disarmament, unilateral nuclear disarmament, the abolition of sovereignty, the creation of a superstate, and, not so long ago, preemptive wars (America against Russia, Russia against China). Others have solved the problem in their own minds by wishing it away; some in this group suggest that the Russians aren't nearly so bad as our leaders say, and if we would only stop provoking them all would be well. Still others suggest that if our leaders would only "stand tall in the saddle" our enemies would be cowed into decent behavior.

Where does the SDI stand in this regard? Is it a technological panacea, mere wishful thinking, or is it a grand idea whose time has come?

From the beginning of the nuclear age until the Reagan administration, all American presidents have placed great emphasis on searching for and exploiting political means for maintaining peace and stability and have played down the possibility of an active defense against a nuclear attack. In addition to maintaining an adequate level of nuclear deterrence, they forged and strengthened alliances, they pursued detente and peaceful co-existence by a variety of diplomatic means, and they sought to negotiate agreements that would first limit and eventually reverse the nuclear arms race among the nuclear powers and prevent its spread to other nations. This has not been a partisan matter: every president from Truman to Carter—four Democrats and three Republicans—placed the same trust in political measures, including arms negotiations, as the favored alternative to the naked threat of nuclear revenge as the means for averting large-scale war. In addition, all of the postwar presidents also supported research and development on a variety of active defense schemes. In the 1940s and early 1950s, primary attention was focused on the construction of air defenses intended to keep long-range bombers away from our shores. In the 1960s, the primary focus was on the development of anti-ballistic missile systems. While the prospects for such systems have waxed and waned in accordance with the technological facts as well as the temper of the times, at no point in the thirty years prior to President Reagan did a president place greater emphasis on developing technological means of defense against nuclear attack than on promoting political means for maintaining international stability and world peace.

It is this balance between the search for technical solutions and the search for political solutions that President Reagan seems intent on changing. We now have the president's challenge to scientists to build a defense so effective that it might eventually "make nuclear weapons impotent and obsolete," a challenge accompanied by an extraordinary national program administered by a specially created new organization. We also have a long list of statements by the president and some of his closest advisers and political associates concerning the perfidy of the Soviet leaders, and the consequent difficulty of making any useful agreements about important matters with them. While this has not prevented the two countries from resuming arms-control negotiations in March 1985, the complexity of combining negotiations on strategic offensive and defensive systems makes the outcome of these talks very problematic.

Is there reason to believe this approach would be better than the one favored by other presidents? Is there reason to hope that we can, at last, find security through technology? The unhappy fact is that, thus far, whenever we (or others) have sought to solve our national security dilemma by technical means, we have in the end only made matters worse. Back in the 1940s and 1950s, the answer to active air defenses was the introduction of ballistic missiles. Ballistic missiles overflew the air defense, making the attack more certain and much quicker. In response to ABM defenses, efforts in the early 1960s to put multiple warheads on missiles (MIRV) were accelerated. This increase in warheads could saturate and swamp any conceivable defense, at the same time making it possible to increase greatly the number of individual targets that could be attacked. MIRVs, however, make the world more dangerous because, when combined with increasing accuracy, they make a first strike seem more promising, thus undermining stability in certain crises conditions.

It is easy to understand why we look to high technology for solutions to our military problems. The United states, after all, is the world leader in this area; technology truly is one of our strongest suits. Those who support various extreme "high-tech" ideas, such as the SDI, can and do point to any of a number of remarkable and dramatic past successes. Radar did tip the balance in the Battle of Britain, and nuclear bombs did end World War II earlier than would otherwise have been the case. In 1961, a president issued an order to go to the moon within the decade and we did precisely that. On an everyday level, our homes and lives are filled with "technological miracles," including televisions that pick up signals relayed by satellites from around the world and home computers as powerful as the largest of their institutional ancestors only a generation ago. The computer revolution, after thirty years of surpassing most short-term predictions, seems destined to offer equally rapid rates of change in the years ahead.

Against this background, the Fletcher panel's optimistic extrapolations of current capabilities are not prima facie unreasonable. On the other hand, we must note that most of these frequently cited technological triumphs involved a struggle of man against nature; the SDI, while it does require some new inventions, also involves a struggle of man against man—of measure against countermeasure—and that is quite a different matter.

We are truly on the horns of a dilemma. Is the SDI, as some think, an instance of exceedingly expensive technological exuberance sold privately to an uninformed leadership by a tiny in-group of especially privileged advisors? If it is, then it is a perfect example of what President Eisenhower had in mind when he warned that "in holding scientific research and discovery in respect, as we should, we must also be alert to the equal and opposite danger that public policy could itself become the captive of a scientific technological elite."

Or is the SDI, as others maintain, a good idea whose time has come? Is it a possible means for replacing assured destruction by assured survival, even, perhaps, for eventually making nuclear weapons obsolete? The issues involved in the debate over strategic defenses are many and complex, and the debate itself will surely go on for many years. What will be vitally important during this debate is that the American public realize that the SDI is not proceeding in a vacuum, but is enmeshed in a web of issues concerning defenses against all types of nuclear weapons, as well as current and projected military uses of space, anti-satellite weapons, offensive nuclear forces, and strategic stability between the superpowers.

References

1. Preliminary Design of an Experimental World-Circling Spaceship, (Santa Monica: RAND Corporation, 1946); see especially the section on the possible military applications of satellites by Louis Ridenour, "Significance of a Satellite Vehicle."
2. Dr. Robert Cooper (director of the Defense Department Advanced Research Projects Agency), The San Diego Union, June 17, 1984, page C4.
3. See the contribution to this volume by Paul Stares, "U.S. and Soviet Military Space Programs: A Comparative Assessment," on p. 127.
4. Speaking about the ocean reconnaissance systems specifically, Dr. Cooper noted that the Soviets have completed a number of experiments designed eventually to provide "over the horizon targeting" information directly from satellites to missiles launched from submarines and aircraft. Given the ability of both aircraft and missiles to achieve much greater accuracy by using satellite data for in-flight navigation, the military need for ASAT weapons to destroy such satellites will become stronger. See The San Diego Union, June 17, 1984, page C4.
5. For the unclassified summary of the Fletcher panel report, see "The Strategic Defense Initiative: Defensive Technologies Study" (Washington, D.C.: Department of Defense, March 1984).
6. James C. Fletcher, "The Strategic Defense Initiative," testimony before the Subcommittee on Research and Development, Committee on Armed Services, U.S. House of Representatives, March 1, 1984.
7. Sidney D. Drell, Philip J. Farley, and David Holloway, The Reagan Strategic Defense Initiative: A Technical, Political and Arms Control Assessment (Stanford: Stanford University, 1984).
8. See especially Ashton B. Carter, Directed Energy Missile Defense in Space, Office of Technology Assessment Background Paper (Washington, D.C.: U.S. Congress, 1984) and John Tirman, ed., The Fallacy of Star Wars (New York: Vintage Books, 1984).
9. The New York Times, March 30, 1983, p. A14.

Why SDI?

WITH SANFORD LAKOFF

Since the onset of the Cold War, and particularly after both superpowers began to amass large arsenals of nuclear weapons, military
planners in the East and West have encouraged efforts to develop
defenses against nuclear attacks. Both sides have made effective use of
"passive" defenses, such as the hardening and dispersal of weapons systems likely to be the prime targets of a preemptive strike and the provision
of shelters for command authorities and vital communications centers.
Both sides have also tried, but with far less success, to develop "active"
defenses, but against the varied and daunting challenge of modem strategic systems, all such efforts so far have been largely in vain, even though
they have been undertaken at considerable expense and with great technical sophistication. The main reason is that every advance in active defense
has been offset by compensatory improvements in offensive forces.

The futility of the effort to build defenses effective against missiles was
in effect acknowledged by both superpowers when they agreed to the
ABM Treaty of 1972. The treaty limited the two countries to no more than
100 ground-based ABM launchers, each with a single warhead, and certain radar detectors and trackers, to be deployed either around a missile
field or the national capital. This severe restriction on ballistic missile defense (BMD) appealed to leaders on both sides because, in the first place,
they were compelled to recognize that the technology for an effective defense simply was not available. As former Secretary of Defense Robert S.
McNamara said, in a remarkable address in 1967 which helped pave the
way for the treaty, "While we have substantially improved our technology
in the field, it is important to understand that none of the systems at present or foreseeable state-of-the-art would provide an impenetrable shield

over the United States."[1] In addition, both parties recognized that limiting defenses would reduce the incentive to add or strengthen offensive forces. The terms of the treaty, however, did not prohibit improvement of existing ABM systems or research that might lead to the development of different defensive systems based on "other physical principles," such as lasers, that were not incorporated into ABM systems actually in use when the treaty was signed. Such research continued to be performed both in the U.S. and U.S.S.R., but there was no substantial agitation on the part of the technical communities in either country for a reconsideration of the technical premise on which the ABM Treaty had been based.

It therefore came as a considerable surprise, even to the technical communities, when in a television address on 23 March 1983, President Ronald Reagan announced that the U.S. was embarking on a major new effort, subsequently called "the Strategic Defense Initiative," to determine whether effective defenses could be built against nuclear attack. How did this dramatic change come about? What were the underlying motives behind the decision? Was it all but inevitable, given the relentless advance of technology, or was it more the product of other considerations— political, strategic, moral, and economic?

SDI, the evidence suggests, was far from inevitable. Unlike virtually all other comparable efforts of weapons innovation since World War II, this one reflected nothing so much as the mindset of a single person, the president who enunciated it on the recommendation of a few likeminded political supporters. The decision was adopted without benefit of prior review by specialists in defense technology and strategy. It was not considered in the president's formal cabinet or by leaders of Congress or U.S. allies. The policy attracted initial support because Reagan succeeded in making a direct appeal to American public opinion. He sensed correctly that the initiative would strike a responsive chord among a majority of voters, especially because it promised to remove the threat of nuclear attack by relying not on the imperfect strategy of retaliatory deterrence or on persuading the Soviets to accept meaningful arms control, but solely on faith in the national capacity for technological innovation. In these peculiar circumstances of its birth, however, may also lie the seeds of SDI's undoing. As the president's popularity has declined, and as his term nears an end, SDI has become more and more vulnerable to attack from all the forces the president tried to bypass. Because it is so closely identified with the Reagan presidency rather than with the coalitional consensus that sustains other major military programs, SDI may not survive his term in office—at least as the high visibility, high priority program he tried to make it.

In examining the grounds for the decision, the history of weapons innovation offers some preliminary guidance, but there is no exact parallel in previous experience, and none of the usual explanations applies well to the case of the SDI. In previous instances, certain factors either have been essential or at least implicated. Two have been especially prominent. Innovation usually has been inspired by fear that an adversary was working on a particular weapon against which a defense was needed or which it was necessary to deter by developing the same weapon. The decision to initiate a program has been reached after experts have judged it likely to succeed. President Roosevelt's order to create the Manhattan Project had both elements behind it. It was made after refugee nuclear physicists persuaded him and his advisers that Nazi Germany was certain to be working on the development of the weapon and might succeed in time for it to affect the outcome of the war.[2] Similarly, President Truman decided to approve a "crash program" to develop the hydrogen or thermonuclear bomb out of fear that if the Soviets developed it first, they might use it to blackmail the West. Although he chose to disregard the advice of a key committee of experts, he acted on the advice of officials who favored the project and with strong encouragement from other prominent scientists; and even those who opposed the project thought it had an even chance of succeeding within five years.[3]

Other factors are also usually in play. Rivalry among the armed services, "bureaucratic politics," salesmanship on the part of defense contractors, the interest of well-placed congressmen in obtaining procurement contracts for their districts, the findings of strategists and other technical experts, and the general interest of the Defense Department in promoting modernization, have all figured in decisions to develop and deploy new weapons.[4] There is also some reason to suppose that innovation occurs because of an "action-reaction phenomenon," in McNamara's phrase,[5] except that most of the time the United States has been the initiator rather than the reactor. The reason is not that the U.S. is any more aggressive than the U.S.S.R. or less conceded about the dangers nuclear weapons pose, but rather that "we are richer and more powerful, that our science and technology are more dynamic, that we generate more ideas of all kinds."[6] This American penchant for innovation, however, has never been completely autonomous; it has been directed and channeled by a combination of technical and policy judgment.

On all these counts, the SDI is different. The SDI decision was not reached on conventional grounds or in the conventional way. There was no reason to suppose that the Soviets were on the verge of achieving breakthroughs that so far had eluded U.S. researchers or that these breakthroughs would be sufficient to allow for a "break out" from the ABM Treaty that would yield some tangible military advantage. Although Soviet scientists had done important research on various types of lasers and had continued to mount a major effort to exploit space for military purposes, including the development of a co-orbital Anti-Satellite (ASAT) weapon, there was no indication that their progress had brought them to the verge of developing radically new, more effective ballistic missile defenses. In 1977, the magazine *Aviation Week & Space Technology* had carried sensational reports, based on information from a recently retired former Director of Air Force Intelligence, Maj. Gen. George J. Keegan, claiming that the U.S.S.R. was close to achieving an operational particle beam weapon to be used for BMD. These claims were reviewed by an Air Force-CIA intelligence panel and a panel of the Air Force Scientific Advisory Board, and rejected as greatly exaggerated. As more credible reports continued to appear confirming Soviet research on beam weapons, the DOD reviewed its assessment at least once a year after 1977.[7] A review conducted by the Defense Advanced Research Projects Agency (DARPA) a year before the president made his surprise announcement had concluded that there were no technical grounds for supposing that a more advanced BMD system could be developed that could not be successfully countered and would not be more expensive to deploy than the offense needed to defeat it.[8] There is little if any evidence indicating that military contractors were actively lobbying to promote a new venture in defense, even though in retrospect it has been pointed out that they might well look to SDI to pick up the slack left as major offensive weapons reached maturity. The very few advisers whose opinions were solicited by the president were almost all predisposed in favor of it, and certainly could not be said to represent a consensus within the technical community.[9] Administration spokesmen later sought to make a virtue of this neglect of expert opinion by citing past instances in which eminent scientific authorities had been proven unduly pessimistic.

More than any previous major decision on weapons innovation, SDI was very much a presidential decision. It was, as experienced observers were quick to note, a "top down" decision rather than one reached, as most had been in the past, after prolonged gestation at various levels of the defense establishment and review by expert committees. From the Truman administration onward, these decisions have generally risen through the upper layers of bureaucratic and advisory network until they have been given an official imprimatur in the form of a National Security Council directive, beginning with NSC 68 in 1950, which was the policy basis for a major increase in defense spending to meet a perceived Soviet threat. When President Eisenhower decided to approve the development of continental air defenses and to accelerate development of ICBMs, it was only after elaborate inquiries by committees composed of experts drawn from within and outside the government.[10] SDI is an anomaly. It was a decision reached by the president without prior review by the defense establishment, in the knowledge that such a review would have recommended against the project, on the advice of an informal "kitchen cabinet," composed of political supporters, after the president himself had developed a leaning in the direction of the project. Not until after the 23 March speech and the issuance on the same date of National Security Decision Directive 85, establishing the goal, did the president issue National Security *Study* Directive 6-83 creating the Defense Technologies Study (chaired by James C. Fletcher) and the Future Security Strategy Study (chaired by Fred S. Hoffman) to examine the feasibility and potential impact of the proposals. Reagan himself scoffed at the suggestion that the idea for SDI could not have originated with him: "It kind of amuses me that everybody is so sure I must have heard about it, that I never thought of it myself. The truth is I did."[11]

Although the decision did not come altogether "out of the blue," the events that led to it were almost exclusively political and personal. The 1980 Republican platform had endorsed a hawkish resolution introduced in Congress calling for "peace through strength" and including among other strategic goals the need to pursue "more modern ABM technologies" and "to create a strategic and civil defense which would protect the American people against nuclear war at least as well as the Soviet population is protected."[12] In 1940, as a Hollywood actor, Reagan had played a role in the film "Murder in the Air" about a secret miracle weapon, an "inertia projector," that could bring down aircraft by destroying their electrical systems.[13] Almost four decades later, as a candidate for the presidency in 1979, he had become aware of the complete vulnerability of the U.S. to attack during a campaign visit to the North American Defense Command headquarters at Cheyenne Mountain, Colorado. After observing NORAD radars tracking thousands of objects in space, he asked the commander of the facility, Gen. James Hill, what NORAD could do to stop a Soviet missile once it had been identified. According to Martin Anderson, an aide who accompanied Reagan on the visit, Gen. Hill replied that NORAD could only alert the officials of the city. "That's all we can do. We can't stop it."[14] Later in the campaign, Reagan reviewed what he learned on his visit in an interview with a reporter:

> [W]e can track the missiles if they were fired, we can track them all the way from firing to know their time of arrival and their targets, and we

couldn't do anything to stop the missiles.... They actually are tracking several thousand objects in space, meaning satellites of ours and everyone else's, down to the point that they are tracking a glove lost by an astronaut that's still circling the earth up there. I think the thing that struck me was the irony that here, with this great technology of ours, we can do all of this yet we cannot stop any of the weapons that are coming at us. I don't think there's been a time in history when there wasn't a defense against some kind of thrust, even back in the old-fashioned days when we had coast artillery that would stop invading ships if they came.

He went on to discuss Soviet efforts in civil defense and added, significantly:

I don't know whether we should be doing the same things of that kind but I do think that it is time to turn the expertise that we have in that field—I'm not one—but to turn it loose on what do we need in the line of defense against their weaponry and defend our population, because we can't be sitting here—this could become the vulnerable point for us in the case of an ultimatum.[15]

Impressed by Reagan's reaction, Anderson drafted a memo in August suggesting that Reagan propose construction of a protective shield against missile attack. The memo was passed to Michael K. Deaver, then a senior campaign adviser. Deaver liked the idea in principle, but not the timing of the proposed statement. He was afraid it would make Reagan look like too much of a "hawk" for proposing such a radical change in strategic doctrine.[16]

Although Deaver's advice prevailed and Reagan did not make known his views on strategic defense during the campaign, other evidence indicates that he was already strongly in favor of doing so. Another of his campaign advisers was a retired Army officer, Lt. Gen. Daniel O. Graham, who, although he was not a technologist, had been head of the Defense Intelligence Agency and was already an outspoken champion of the military need to "seize the high ground of space" before the Soviets did. Graham, along with other military men and space enthusiasts ("space cadets" to their critics), was strongly convinced that just as warfare had earlier spread from land to sea and then into the air, so it also was bound to expand into outer space. A later Air Force "Space Master Plan," setting objectives to the end of the century, would call for a "space combat" capacity aimed at protecting Air Force assets and denying the enemy free access to space.[17] For Graham, as for many Air Force officers, the question was not whether the U.S. and the Soviets were in a competition for military control of space, but which side would win. He has recalled that while a candidate, Reagan often objected to the concept of Mutual Assured Destruction, referring to it as a "Mexican standoff." Graham added,

"He said it was like two men with nuclear pistols pointed at each other's heads, and if one man's finger flinches, you're going to get your brains blown out."[18] Reagan, who is known for his tendency to repeat favorite images of this sort, drew a similar analogy in explaining SDI to reporters shortly after he announced it.[19]

After the election, Graham set about building support for his idea among the president's staunchest supporters. Early in 1981, he published an article in *Strategic Review* calling for the deployment of space-based defenses using "off the shelf" technology.[20] Later in that year he founded High Frontier, an organization committed to the view that the U.S. should aim to achieve military superiority over the U.S.S.R. by making a technological "endrun" on the Soviets.[21] Space would be the key arena in which this technological edge would be exploited, at first using kinetic kill technology— systems designed to shoot projectiles or "smart rocks" at missiles, destroying them by impact—and later, as they became available, more advanced systems incorporating beam weapons. By taking this bold approach, the advocates of High Frontier contended, the U.S. would not have to be content merely with restoring a preexisting balance in offensive systems, but could achieve superiority by adding defenses to offenses. The cost of deploying the proposed system was estimated as "on the order of $24 billion" in constant 1982 dollars. The ultimate aim would be to replace MAD by "assured survival."[22] Among the contributors to High Frontier were four members of the president's California kitchen cabinet: brewer Joseph Coors, retired industrialist Karl R. Bendetsen, investor and longtime Reagan friend William A. Wilson, and food magnate Jaqueline Hume.

At the urging of Reagan appointees sympathetic to High Frontier, its proposal of a three-layer Global Ballistic Missile Defense (GBMD), consisting of terminal, mid-course and boost-phase systems, was examined by specialists in the Defense Advanced Research Projects Agency (DARPA), the agency of the Department of Defense in charge of research in the general area of future technologies. The agency found that the proposed system would be ineffective against the present Soviet missile force, vulnerable to countermeasures, and considerably more expensive than the organization claimed. In testimony to Congress, DARPA Director Robert S. Cooper reported on the agency's findings, noting that its researchers had worked with representatives of High Frontier:

> [W]e do not share their optimism in being able to develop and field such a capability within their timeframe and cost projections. We have conducted several in-house analyses and have experienced some difficulties in ratifying the existence of "off the-shelf components or technologies" to provide

the required surveillance, command and control, and actually perform the intercepts within the orbital and physical conditions described. Our understanding of the system's implications and costs would lead us to project expenditures on the order of $200 to $300 billion in acquisition costs alone for the proposed system.[23]

On the recommendation of the General Accounting Office,[24] however, the DOD in 1982 set new priorities for the research on strategic defenses and put the relevant activities under the guidance of DARPA.[25]

Also promoting strategic defenses was a "laser lobby" in Congress headed by Republican Senator Malcolm Wallop of Wyoming and his legislative assistant, Angelo Codevilla. In 1979, both men drafted and sent to Reagan an article they had written calling for immediate deployment of space-based defenses featuring chemical lasers. Reagan returned the draft with comments and annotations.[26] At Wallop's urging, the Republican-controlled Senate voted in 1982 to provide additional funding for laser defenses by a margin of 91–3. In the House there was some enthusiasm for the X-ray laser, as *Science* magazine reported in a story entitled "Laser Wars on Capitol Hill" in its issue of 4 June 1982. Another Republican Senator, Pete Domenici of New Mexico, had published an article in the conservatively inclined *Strategic Review* calling for a new program to develop space-based defenses.[27] But there were countervailing pressures. In 1982, both houses voted to deny $350 million requested by the Defense Department for research on ABM systems, and a joint resolution was introduced in the House of Representatives, signed by ninety congressmen, calling on Reagan to negotiate a ban on all space weapons; a similar resolution was introduced in the Senate.

Although the specialists at DARPA were clearly unimpressed by the case for a high-priority program in strategic defense, the cause received influential support from the nuclear physicist Edward Teller. A former director, and at the time a kind of physicist-in-residence at Lawrence Livermore Laboratory, Teller had been encouraging a group of proteges—many of them brought to the laboratory under a fellowship program of the Hertz Foundation, in which Teller long has played a key role[28]—to investigate a "third-generation" nuclear device (incorporating fission and fusion weapons but focusing the energy of their explosions in a powerful beam of X-rays) which might have military applications. In 1967, Teller had met Reagan, when he was Governor of California, and had given him a tour of Livermore Laboratory.[29]

Two younger admirers of Teller have also played key roles in promoting SDI. The leader of the "Excalibur" project team investigating the X-ray laser at Lawrence Livermore National Laboratory, dubbed "O Group," was Lowell Wood, a younger physicist who later would appear often with Teller in Congressional hearings and otherwise played an active role in lobbying for support of SDI in general and the X-ray laser in particular. George ("Jay") Keyworth II, then head of the physics division at the Los Alamos National Laboratory, and also a protege of Teller, became a strong advocate of Teller's views in general, even though he did not share Teller's enthusiasm for the X-ray laser, after being appointed Special Assistant to the President for Science and Technology in 1981.

Teller long had been in favor of pursuing passive defenses in the form of civil defense, including the dispersion of population and industry, in order to limit the damage from a nuclear attack. Now he pressed Keyworth to promote an effort to develop active defenses, including research on third-generation nuclear weapons. With Keyworth's encouragement and Teller's active participation, the four members of the kitchen cabinet reportedly decided to act separately from High Frontier. They conferred at the offices of the conservative Heritage Foundation in Washington and used their considerable influence to arrange a White House meeting with Reagan in January 1982 in which Teller took part. At that meeting they presented a report urging the president to establish a strategic defense program modeled on the World War II Manhattan Project. On 14 September 1982, Teller met separately with the president and key advisers. Although he had been invited for a different purpose—Keyworth wanted his help in persuading the president to increase support for basic research—Teller used the opportunity (much to his protégé's annoyance) to urge increased support for the X-ray laser project. He is reported to have warned the president that the Soviets were making significant progress in developing the new laser and to have advised him that a major breakthrough in the same effort had been achieved in the "Excalibur" project at Livermore.[30] Teller has said subsequently that "because the Soviets are doing it, by now it is a question of life and death," and that the achievement of the X-ray laser "would end the MAD era and commence a period of assured survival on terms favorable to the Western alliance."[31]

Teller's promotion of the Livermore project did not reflect a belief that the U.S. should deploy defenses only when the X-ray laser had been perfected, or that it should become committed to a space-based defensive system, either using kinetic weapons on satellites like the High Frontier project or also relying on battle stations armed with laser weapons. Indeed, Teller specifically rejected the idea of satellite-based interceptors, and declined to endorse the High Frontier proposals because they made specific use of this basing mode. "We are not talking about battle stations in space," he had told the House Armed Services Committee. "They are much too vulnerable. We should merely try to have our eyes in space and to maintain them."[32] Rather, Teller's support of strategic defense using new technology was general rather than specific, and he had long opposed any political actions—including treaties—which might hamper the development or deployment of defensive systems. In a book published in 1962, Teller already was campaigning for active defenses. "A retaliatory force," he wrote, "is important. A truly effective defense system would be even more desirable. It would be wonderful if we could shoot down approaching missiles before they could destroy a target in the United States." If the Soviets were to develop reliable defenses knowing that U.S. defenses were insufficient, "Soviet conquest of the world would be inevitable."[33] He had been opposed to the 1963 Limited Test Ban Treaty because it prevented development of nuclear-tipped ABMs, and to the ABM Treaty because it prevented the U.S. from deploying more than a small number of ground-based interceptors. Now, he continued to believe that such interceptors would be worth having, especially in view of advances that had been made in their design, even if they could not promise complete protection and even though, in the long-run, the X-ray laser, if it could be popped up from submarines on warning of an attack, would provide a much more effective defense, by intercepting missiles in their boost phase without the vulnerability of space-based satellites.

In his meetings with the president, Teller evidently emphasized the long-term prospects opened by the X-ray laser, though he very likely also reiterated his long-held belief that some defense is better than none at all. Later, Teller was to react strongly against the conclusions of the Fletcher committee, appointed by the president after his Star Wars speech to propose a way of implementing the president's vision. Teller objected because the Fletcher report called for deferring deployment until research into all possible alternatives had been completed, and because the report anticipated the deployment of a layered defense requiring some space-based systems. "The spirit is willing," Teller punned, "but the Fletch is weak."[34]

Teller's technical judgment was given practical application as a result of Reagan's meetings involving his kitchen cabinet, Graham, and a small group of White House officials, none of whom had major responsibility for foreign or defense policy. On 8 January 1982, the outsiders presented their recommendations to the president. During the meeting, the question arose of whether an ABM system should be designed to rely on beam weapons and whether the system should be expected to provide an area defense to protect people or a "point defense" that would protect only missile silos and other military targets. Bendetsen reportedly suggested that if one type of defense could be achieved, so could the other. The real need, he said, was to get on with the job. The president agreed. Little thought was given to the question of whether the proposed defenses would be compatible with the obligations imposed by the ABM Treaty. The right way to proceed, the participants agreed, was for the president to issue a directive initiating a program on the scale of the Manhattan Project. For something that elaborate, Reagan observed, he would need a recommendation from the DOD. Bendetsen, who had been Under Secretary of the Army from 1946 to 1952, observed that if that were the case, Reagan would have to talk the department into it.[35]

The others at the White House meeting recognized the force of Bendetsen's point. They knew that if the president were to try to win approval through the usual channels, the proposal would run into so much opposition that it might never reach his TelePrompter. Within the defense establishment, the opposition was likely to have been keen, especially among those who had already examined the High Frontier proposal and concluded that it was premature at best. Leaders of the various services were apt to fear that large appropriations for strategic defense might jeopardize future modernization of offensive weaponry. Those most knowledgeable about advanced technologies and the costs of development would be either skeptical or strongly opposed. On the very day the president made his "Star Wars" speech, the director of the DOD's Directed Energy Program told a Senate committee that while beam weapons "offer promise of making major contributions" to strategic defense, the "relative immaturity" of the technologies made it hard to know whether weapons employing them would be feasible or cost-effective. For the time being, he indicated, "our goals in this area are rather modest." The research would be unlikely to affect force structure until "the 1990s or beyond."[36] Richard DeLauer, Under Secretary of Defense and in effect the highest ranking technologist in the Pentagon, said later in the year that a deployable space-based defense was at least two decades away and would require "staggering" costs. To develop it, he added, eight technical problems would need to be solved, each of which was as challenging as the Manhattan Project or the Apollo Project.[37] And even if it could be developed, DeLauer was skeptical of its utility in the face of countermeasures: "There's no way an enemy can't overwhelm your defense if he wants to badly enough."[38]

Nor would the proposal have been any more welcome elsewhere in the executive branch or among most congressmen. Had it been circulated in advance, the State Department certainly would have warned that reaction in Western Europe would be hostile and that Soviet objections might undo the ABM Treaty. Arms-control specialists in the department and in the

Arms Control and Disarmament Agency might have cautioned that it would complicate arms reduction negotiations with the Soviets. And had the proposal leaked to Congress, those already skeptical of the proposals of High Frontier and the laser lobby might well have mounted a campaign to head it off.

Once the president decided to commit himself to the program, he therefore shared his decision only with a small circle of National Security Council staff members, headed by deputy national security adviser Robert C. McFarlane. The basic strategy adopted, according to McFarlane, "was to skirt Congress, the bureaucracy, and the media." Secretary of State George P. Shultz was not consulted. He was simply given an advance copy of the speech, two days beforehand, which was marked "eyes only," preventing him from sharing it with arms control adviser Paul Nitze, who first learned about the project on the day the speech was given. DeLauer learned of the decision nine hours before the speech. Fred C. Ikle, Undersecretary of Defense for Policy, was notified at the same time and pleaded for an opportunity to inform at least the leaders of the NATO allies.[39] The process by which the policy was formulated was similar to that later followed in the Iran Contra affair. Here too, the president relied primarily on the staff of the National Security Council rather than on members of his formal cabinet and the relevant executive agencies.

The president's science adviser, who had been privy to the earlier discussions with Teller, was brought in only after the president had decided to act. He was informed five days before the date of the speech, when the NSC staff realized that the president could not go on television to announce a major high-technology initiative without telling his science adviser anything about it. Keyworth has said that the president asked him to determine whether the objective was attainable, and that he spent several days telephoning experts before he was able to give the president the assurance he was hoping to get. He had admitted that his first reaction was to ask for time to consider the issue: "My God, let's think about this some more. Let's think about the implications for the allies. Let's think about what the Soviets are going to think. Let's think about what's technically feasible. Let's think about what the scientists are going to say. Let's think about the command and control problems."[40]

Keyworth's initial concerns were reenforced when he consulted a specialist on directed energy weapons, but in the end he overcame his doubts, largely out of loyalty to the president. In 1981, he had commissioned an independent review of prospects for space-based defense by a physicist, Victor H. Reis, whom he had appointed assistant director of the Office of Science and Technology Policy. Reis, who had worked at the MIT Lincoln Laboratory and had participated in Defense Science Board panels on strategic defense, reported that it would be much easier to counter such a defense than to build it. When Keyworth was given a draft of the president's "Star Wars" speech by McFarlane, he showed it to Reis, whose reaction was to say, "Jay, this is Laetrile." Keyworth admitted that he too had doubts about the idea, but thought that SDI might be a boon to scientific research. Reis suggested that he should either resign or urge the president to submit the proposal for review by a panel of independent experts. Keyworth knew that the Joint Chiefs also favored the appointment of an independent panel, like the panel headed by the physicist Charles Townes that had considered the MX. Keyworth rejected the advice because he had confidence in Reagan's political leadership and because he thought of SDI as a research program, not necessarily implying a commitment to deploy any particular system. Above all, he decided to endorse the proposal because he thought that the proposal was a political decision and that his role as science adviser was merely to make sure that the president was aware of the views of the technical community. He also was influenced, however, by previous finding indicating that a laser defense might be developed which would rely on ground-based lasers and require only that mirrors be deployed in space. Once he had decided to go along with the proposal, he helped McFarlane draft the speech.[41]

Several members of the White House Science Council were invited to attend the broadcast of the speech, but they too were taken by surprise. A year before, a panel of experts on the council had been asked by Keyworth to study the potential impact of new technologies and had reported that none of them, including the X-ray laser, was likely to have a revolutionary impact in the near term. Many of the members of the council were nevertheless sympathetic to the president's proposal. Only one member, John Bardeen, professor of physics and double Nobel laureate at the University of Illinois, dissented strongly. He resigned from the council in April because "such a questionable and far-reaching proposal was being made without review by the scientific community." Bardeen's complaint emphasized the lack of advanced review by technical specialists:

> Reagan's speech was prepared without prior study of the feasibility of the concept by technical experts and apparently without consultation with those in the Pentagon concerned with missile defense or with his own former science advisor, George "Jay" Keyworth, Jr. I was a member of the White House Science Council at the time. Although we met only a few days before the speech was given, and had a panel looking into some of the technology, we were not consulted. We met on a Friday and left for home. Keyworth must have heard about the planned speech shortly thereafter, be-

cause telegrams were sent to some of the council members involved to return to Washington on Saturday. With the speech scheduled for the following Wednesday, there was no time to make more than minor changes. The multi-layer systems on which current SDI research is based were outlined in a report by the Fletcher Commission, formed after the speech was given.[42]

The lack of consultation was quite deliberate. The president's reasoning, which was reenforced by McFarlane, was that a decision needed to be made which would change attitudes toward defense both within the government and outside it. "The idea," as one of the participants put it, "was to make a decision and then make it happen." That way, no one had to consider whether a defensive system might work or whether, assuming it could be made to work, there would be military benefits. Nor was there any thought given as to whether the Soviets would find the announcement threatening or whether it might have any effect, adverse or positive, on existing treaties and the prospect of new arms control agreements. As a White House source insisted,

> None of the things you assume would be considered were considered at all. People just don't believe that the president could make such a momentous decision so impulsively. They think we must have thought through what it could do to the treaties and how it might work as a bargaining chip in Geneva, and so on, but I can't find it.

Another participant said the decision had been given little more consideration than "you would give the jacks before you bounce the ball." Still another gave a post hoc appraisal implying more admiration for the way the decision had been reached: "It was a fabulous study of top-down leadership." He noted that no one the president informed of the decision had objected that it was "screwy."[43]

Given the disposition of those who were consulted in advance, it is perhaps not surprising that the president encountered no criticism. Most of the top civilians in the administration with foreign policy responsibility shared his view that the ABM Treaty had been a mistake, because it had discouraged the U.S. from proceeding with research on defenses while not inhibiting the Soviets, and had not led to the promised reductions in offensive arms. In addition to Graham and Teller, and such politicians as Wallop and Representative Newt Gingrich, conservative strategists such as Colin Gray and William van Cleave had condemned the treaty and were urging a renewal of defensive research.

Nevertheless, there were objections to details of the proposal, and there was confusion among the drafters as to how it should be couched. As a result, the speech was shaped to take some account of friendly criticisms and concerns. The drafters worried whether the president should be circumspect or forthright—whether he should approach the subject by saying something like "it might be a good idea to think about..." or whether "As of tonight, I have directed...." Richard Perle, Assistant Secretary of Defense for International Security Policy, succeeded in narrowing the focus to protection against ballistic missile attack, cautioning that to include defense against bombers and cruise missiles would dramatically increase cost estimates for the U.S., and even more for Western Europe. Both Perle and Shultz found the initial draft of the speech unnecessarily provocative toward the Soviets. They foresaw that the Soviets would object that by coupling a defense with a potent offense, the U.S. would put itself in a position to launch a first strike against the U.S.S.R. and absorb retaliation by what remained of the Soviet missile force. The president tried to meet the objection by acknowledging the possible response and attempting to allay it. ("If paired with offensive systems, they can be viewed as fostering an aggressive policy, and no one wants that.") Perle also pointed out that U.S. allies might conclude that the U.S. intended to protect itself but not them. They would fear that a U.S. defense against ICBMs would enhance the likelihood that the U.S. would become "decoupled" from the defense of Europe. The drafters therefore added the clause, "recognizing the need for closer consultation with our allies," and the assurance that the U.S. sought a defense that would also protect them.[44]

The context is also noteworthy. The reference to strategic defense was added to a speech originally designed to rescue the president's defense budget request in the face of efforts in Congress to pare it down. The call for research on strategic defense may well have been designed to capture public interest and improve the political climate for the passage of the president's request for a ten percent increase in the defense budget. But the reference was neither a hastily added afterthought nor something tacked on for publicity value. It was a major new initiative intended to move strategic doctrine and defense technology in a new direction.

In the speech, the president took care to cite one important source of support for his ideas—the Joint Chiefs of Staff. Twice the president referred to the JCS, claiming that he had arrived at his decision only "after careful consultation" with his advisers, including the JCS. In fact, the defense issue surfaced at a JCS meeting mainly by coincidence, because the chiefs were searching for something new and different to discuss with the president. They were holding monthly meetings which were going well, but had in effect run out of new things to raise with him. They did not want to tell him, one participant recalled, just that "readiness was up"; he added, "we wanted to bring the president something new, different and exciting." At a preliminary meeting, Adm. James Watkins, Chief of Naval Operations, proposed that they raise the question of the possibility that new technologies might provide a defense against ICBM attack. He did not have in mind urging the president to launch a major program to develop such weapons. His idea was rather that the United States might pursue such weapons gradually, after consultation with its NATO allies. The chiefs agreed to put the matter on the agenda.[45]

At the meeting on February 11, the issue was raised in the context of a discussion of the politically embarrassing opposition voiced by the Joint Chiefs in congressional testimony concerning the administration's Dense Pack basing proposal for the MX. The chiefs may have been keen on finding another issue on which they could reestablish rapport with the president. Strategic defense was just such an issue. Watkins, a Roman Catholic, had been made especially sensitive to the moral issues surrounding a policy of deterrence by threat of retaliation by his involvement in the church's discussion of the ethics of defense policy. The American Catholic bishops had invited discussion of a draft pastoral letter, criticizing retaliatory deterrence as immoral. The letter contended that "collateral damage," even from a counterforce strike, would inevitably take the lives of millions of innocent civilians and that a limited nuclear attack would be likely to escalate to an all-out exchange. "Wouldn't it be better," Watkins asked, "to save lives rather than to avenge them?"—a question the president thought striking and borrowed for his speech. (Others in the administration, including National Security Advisor William Clark and his successor, Robert McFarlane, may also have been influenced by the discussion provoked by the draft of the bishops' letter.)

Just how enthusiastic the chiefs were about a new initiative in strategic defense is in dispute. Administration spokesmen claim that they were polled on the feasibility of strategic defense and concurred, and the president drew his final determination to launch the program from their counsel. According to the reporter Frank Greve, the ideas expressed by the chiefs were "vague and philosophical." Gen. John Vessey, Jr., the chairman of the JCS, and Gen. E. C. Meyer, the Army Chief of Staff, both of whom took part in the meeting, could not recall being polled. Vessey said afterward that it was well recognized that strategic defense was no panacea, and that the chiefs were aiming merely to urge the president to recognize that the concept deserved further study.[46]

Five years later, in a congressional hearing, McFarlane, who had been present at the February 1983 meeting, confirmed that the president had asked the chiefs whether they were endorsing an effort "to determine... whether defense could make a bigger contribution to our military strategy "against nuclear conflict." According to McFarlane, each of the chiefs indicated his agreement, but in a dramatic colloquy with Congressman Les Aspin (D. Wisconsin), chairman of the House Armed Services Committee, he revealed that the chiefs had not understood their endorsement to apply to population defenses but only to the protection of military assets, and that the president had afterward knowingly altered the proposal to give it that much broader goal:

> CHAIRMAN. But at that point, you and they were talking about defending missiles, not the Astrodome protection of the population. Is that what you are saying?
> MR. MCFARLANE. The military was talking about that, I was talking about that, and also frankly about just stressing the Soviet system.
> CHAIRMAN. Complicating their lives.
> MR. MCFARLANE. That is right.
> CHAIRMAN. How did that get transformed into an Astrodome protecting population by the time the President made his speech about a month later?
> MR. MCFARLANE. Basically, the President wanted to change it, Mr. Chairman.... He made the point to me, and I think it was proper, he said, "You must deal and have dealt with the traditional threat that we have faced, and you deal in military terms with the military problem and military risks, and to a certain extent political risks." He said, "My job is to lead and to try to evaluate what may be feasible technologically, but to be responsive to human beings, and my responsibility requires that I try to physically protect them and to move away from a strategy of threatening to kill people."[47]

It is apparent from the president's own account that he seized the opportunity presented by the chiefs' expression of interest to make it appear that the idea had originated in his meeting with them. His own recollection, as he vividly recounted it to *US News and World Report*, was rather different from Vessey's but close to McFarlane's:

> I brought up the question that nuclear weapons were the first weapons in the history of man that had not led to the creation of a defense system to protect against them. I asked if it was worthwhile looking into this. Is it possible to come up with a defense? They were all agreed it was. And right there the program was given birth.[48]

From that point on, the chiefs were not involved in the actual shaping of the decision and the program it would call for. McFarlane assigned three staff members of the National Security Council, all senior Air Force officers, to think about what should be done, to propose various options, and to consider the implications for planned military systems and the possible international ramifications. Then came the actual drafting and redrafting of the speech.

Most of the president's televised speech of 23 March was devoted to urging support of his defense budget request, but he introduced his remarks by saying that he would announce "a decision which offers new hope for our children in the 21st century." He began by noting that until this point the U.S. had managed to thwart the danger of a Soviet attack by threatening deadly retaliation. This approach, he admitted, had worked for more than three decades. It would continue to work only if the U.S. countered the Soviet effort over twenty years to accumulate "enormous military might." He showed photographs of an extensive Soviet intelligence system in Cuba and Soviet military hardware in Central America, including Grenada. Finally, he broached his surprise announcement, making sure to refer to his discussions with the JCS: "In recent months, however, my advisors, including in particular the Joint Chiefs of Staff, have underscored the necessity to break out of a future that relies solely on offensive retaliation for our security."[49] This initial reference to the need to develop defenses begins, interestingly enough, in a recognition of a potential American vulnerability in the offensive competition with the Soviets—a consideration which in many ways underlines and colors the character of the initiative.

The concern for strategic defense, it is plausible to assume from this reference, did not arise only because of the attractions of defense, but also out of fear that the U.S. was losing ground in the offensive competition with the Soviets. The president and his military advisers were evidently convinced that while the Soviets were increasing their deployments of offensive weapons, and were likely to continue doing so, the U.S. was handicapped in matching Soviet efforts, mainly by political objections at home and abroad, but also by resistance to military spending increases due to reluctance to incur further increases in the national debt. The administration's proposal for the deployment of 200 mobile MX launchers had been stymied by resistance to the new Dense Pack basing proposal. In conjunction with this intended deployment, the administration's proposed fiscal 1984 defense budget included a request for substantial growth in support for the development of new ABM technology to be installed to protect the MX missiles. The objections included protests from inhabitants of the western (and strongly Republican) states in which the missiles would be based, who were concerned, among other things, with the impact of the deployment on land and water resources. Only 50 MX missiles (which the administration decided to call "Peacekeepers") were approved, and these were to be housed in existing silos. Such a deployment hardly offset the Soviet advantage in the throw-weight of its heavy land-based missiles—a sore point with administration strategists—nor could it close the theoretical "window of vulnerability": in theory, the ground-based leg of the U.S. triad could still be threatened with destruction in a surprise attack.

Other developments gave ominous signs that other impending administration efforts to make up for perceived Soviet advantages were also in trouble, at the same time as the Soviets were taking new steps that would widen the gap even further. A resolution calling for a freeze on nuclear weapons at current levels (which meant, in the administration view, that the Soviets would retain their advantages) had been passed in referenda in eight states, and there was growing support for the freeze in Congress. Several influential senators, Republicans Charles Percy and William Cohen and Democrat Sam Nunn, were supporting, as an alternative to the freeze, a "guaranteed mutual build-down" which the administration also disliked. Congressional resistance to the development of ASATs was growing, even though defense officials considered them vital in view of the Soviet development and deployment of satellites that would provide "real time" targeting information, greatly improving the Soviet offensive threat. This prospect was considered especially serious by naval officers responsible for the defense of the surface fleet. While they remained confident that the ability to maneuver would be sufficient to protect against a small number of targeting satellites, they thought that a large deployment of such Soviet satellites would require attack by ASATs if the fleet was to be protected in time of war. For their part, the Soviets had shown no compunctions about testing a co-orbital ASAT, judged to be useful but not very threatening, because of its limitation to low orbits as well as its unreliability and relatively lengthy attack time-line. Because of these tests, the Soviets could be said to possess an operational ASAT, despite their rhetorical attacks on the U.S. for "militarizing space" and their offer to ban all space weapons—an offer interpreted by the administration as an effort to preserve their advantage by halting the American program, which promised to produce a better ASAT. The Soviets were also beginning to test and deploy new mobile missiles, which the administration's defense analysts feared would be harder to target than fixed launchers, especially because they could be hidden in the vast reaches of the Soviet Union or in tunnels. The U.S. Midgetman program, while it was the favorite of some Democratic Congressmen, notably Senator Albert Gore and a number of senior defense analysts, who preferred it to the MX, was mired in controversy within the administration. Although the Scowcroft Commission endorsed Midgetman, critics in the administration feared its approval would end prospects for Congressional agreement to further MX deployment and that its single-warhead configuration would further disadvantage the U.S. They also argued that the cost of Midgetman and the difficulty of locating it so as to overcome political objections would make it unpopular.

Efforts to improve NATO's force structure also were meeting objections in Europe. The deployment of longer intermediate range missiles, requested during the Carter administration by European governments with the prompting of U.S. defense officials, in part as a response to Soviet deployments of the SS-20s against targets in Western Europe, had aroused an unexpectedly intense storm of protest. Faced with a revival of the movement for unilateral disarmament, leaders of the allied governments sought to distance themselves from the president's militant anticommunism and his calls for Western rearmament. It must have seemed to the president, and presumably to the Joint Chiefs as well, that the Soviets were able to do whatever they wanted in developing and deploying nuclear weapons, while the U.S. faced mounting political opposition to efforts to keep up with them from political forces both at home and abroad. Equally important, the Soviets had long taken strategic defense of their leadership, the general population, and their retaliatory strategic forces quite seriously.

The standard or traditional alternative to building more and more weapons to keep up with the Soviets was to pursue arms-control measures, but the president, like other conservatives, had come to believe that arms control was a hopeless cause. He had denounced the SALT II Treaty as "fatally flawed" and in general spoke of arms-control agreements as successful Soviet efforts to tie America's hands while they took advantage of every treaty loophole and did not stint from cheating when the loopholes were not adequate to their purposes. He had appointed representatives to the arms-control talks in Geneva on whom he could count to resist pressures to negotiate treaties that would not require the Soviets to forego the advantages they had presumably won in previous rounds. Edward L. Rowny, a retired Army general who was named chief negotiator in START, had resigned in protest as soon as the SALT II Treaty had been initialed in Geneva by U.S. and Soviet representatives and had opposed its ratification. Paul Nitze, who was appointed chief negotiator in the INF talks, was one of the leaders of the Committee on the Present Danger, which spearheaded the campaign against ratification of the SALT II Treaty, although he was viewed with suspicion by some in the administration because he was a Democrat and had been a SALT I negotiator. The president showed no sign of disappointment when the Soviets walked out of the negotiations in 1983 after the U.S. and its European allies rejected a Soviet demand that deployment of Tomahawk and Pershing II missiles be halted for the duration of the talks. On the advice of the Joint Chiefs, he was adhering to the Salt II limits for the time being, so long as the Soviets also adhered to them. The chiefs agreed with the CIA that the Soviets could add more warheads to their heavy missiles than the U.S. could add to its Minuteman missiles if the treaty limits were abrogated.[50]

The president's negative assessment of arms control was shared by some specialists. The trouble with arms control, Helmut Sonnenfeldt has observed, is that "what is negotiable is not significant, and what is significant is not negotiable."[51] Thomas Schelling wrote that arms control "has gone off the tracks." The insistence on formal agreements, he contended, may actually prevent arms reductions. Reciprocal restraints are more likely to come as a result of tacit behavioral signals than formal negotiations.[52] Others have taken a position even closer to the view of the president and other conservatives by arguing that the Soviets are more likely to respect the power of the Western alliance than they are efforts to conciliate them by cutting back military forces, and that the Soviets cannot be trusted to adhere to arms-control agreements.

In his 1983 speech, however, the president went beyond such standard questioning of the utility of arms control to challenge the very premise of the doctrine of deterrence by threat of retaliation. Even success in arms control would not remove the threat of nuclear war. The security of the United States and its allies would continue to rest on the threat of retaliation and the whole of Western civilization would still have to live with the fear of nuclear devastation. The president therefore raised Watkins's question and asked rhetorically, "Wouldn't it be better to save lives rather than avenge them?" Although it might have strengthened his case to do so, he did not mention the draft pastoral letter of the Roman Catholic bishops, possibly because, like other conservatives,[53] he found their position defeatist and even leftist. It is plausible to suppose, however, that the bishops' disapproval of retaliatory deterrence had not escaped his notice, and that in calling for a change in strategy he hoped to capitalize on their unhappiness with the status quo. The bishops had urged that the world move toward total nuclear disarmament. The president offered the same goal, but he proposed to get there by a different route, one that would not require frustrating, difficult, and possibly unsuccessful negotiations, but would instead take advantage of Western scientific and technological superiority over the Soviets. The belief that the U.S. could somehow defend itself was not altogether new, but never before had it been presented so simply, optimistically, and starkly. As to the freeze campaign, Reagan reiterated the administration line:

> [A] freeze now would make us less, not more, secure and would raise, not reduce, the risks of war. It would be largely unverifiable and would seriously undercut our negotiations on arms reduction. It would reward the So-

viets for their massive military buildup while preventing us from modern-izing our aging and increasingly vulnerable forces. With their present margin of superiority, why should they agree to arms reductions knowing that we were prohibited from catching up?[54]

While the speech was being drafted, the president made it emphatically clear that he had in mind more than ground-based defenses that would protect military targets but not population. According to Keyworth, the president told the drafting group: "If there's one thing I do not mean by this, gentlemen, it is some kind of a string of terminal defenses around this country." It was Reagan himself who added the phrase about making nuclear weapons "impotent and obsolete" and insisted on its retention when some of his advisers urged that the phrase to be toned down.[55]

Reagan presented the alterative to reliance on nuclear deterrence in terms at least as appealing as those by the champions of arms control and disarmament. "What if free peoples could live secure," he asked tantalizingly, "in the knowledge that their security did not rest upon the threat of instant retaliation to deter a Soviet attack, but on the knowledge that we could intercept and destroy strategic ballistic missiles before they reached our own soil?"[56] (To listeners who did not stop to think that the president was deliberately not offering the prospect of a comprehensive defense against nuclear attack, by neglecting to call for a defense against the "air-breathing threat" posed by bombers and cruise missiles, it might well have seemed that he was promising total safety from nuclear attack. He said nothing then or later to dispel such an illusion.)

The president did not claim that his goal could be achieved rapidly, but he argued that it was worth pursuing as a matter of the highest national priority. Building such a defense, he pointed out, was a formidable challenge which might not be met before the end of the century, but the goal of rendering strategic nuclear weapons "impotent and obsolete"—a phrase that was to be often repeated, but was to prove more ambiguous than the president and his speech writers may have realized—was both worthwhile and within the realm of possibility. In a pointed reference to weapons researchers, the president called upon those in the scientific and engineering communities who had given the nation nuclear weapons to bend their efforts now to developing a defense against those weapons.[57]

The president tried to fend off one objection that he had been warned about when the draft of the speech was circulated, but his efforts were somewhat perfunctory and they did not prevent the objection from being forcefully advanced. He acknowledged that by deploying a defense along with a strong offense, a nation might be thought to be seeking military superiority, which could be exploited either for purposes of attacking an adversary or for what Schelling had labeled "compellence," i.e., making political capital out of a perceived military superiority and using it to force concessions. "If paired with offensive systems," he noted in the speech, "defensive systems can be viewed as fostering an aggressive policy, and no one wants that." He pledged, "We seek neither military superiority nor political advantage."[58] This assurance did not prevent Soviet Premier Yuri Andropov from promptly denouncing the new initiative as a transparent effort to disarm the Soviet Union by achieving "a first nuclear strike capability," combining offenses and defenses in order to render Soviet forces unable to deal a retaliatory strike.[59] The Soviet argument was cited sympathetically by prominent domestic critics of the president's initiative, who also warned that defenses could be used to absorb a ragged retaliatory strike.[60]

What was mainly on the president's mind in making the speech, however, was apparently not such subtleties and complications but a simple heartfelt concern for a defense against nuclear attack that might serve as substitute for reliance on an offensive threat that could not prevent terrible damage. This concern was expressed with particular force a few days after the speech was given in a press conference: "To look down to an endless future with both of us sitting here with these horrible missiles aimed at each other and the only thing preventing a holocaust is just so long as no one pulls the trigger—this is unthinkable."[61] Presidents before him had expressed the same horror at the thought that they might have no alternative but to seek to defend the nation by actions carrying a distinct risk that a nuclear holocaust would take place. The difference was that whereas they all had reluctantly concluded that there was no feasible alternative to acceptance of this awesome responsibility, he was eager to embrace the hope that one might be found, and to embark on a major program based on little more than that hope. Even more important, he was determined to pursue this objective regardless of its effects on accepted strategic doctrines or the arms-control process.

Although the president arrived at his hope without benefit of a thorough study of the technological feasibility, and although he deliberately sought reassurance on this score from a very limited set of technical advisers who could be counted on to endorse his view, rather than also asking the advice of the far larger constituency of skeptics, his decision was informed—though without real understanding of the complexities involved—by the technical progress that had been made in a number of relevant areas, including the third-generation nuclear weapons touted by Teller and his proteges. The program was labeled "Star Wars" in the media, after the popular film of that title, because it was assumed that Reagan had in mind a space-based defense such as was being advocated by the High Frontier group or one that might involve the orbiting of laser weapons, even though his speech itself contained no specific reference to space-based systems or to any particular defensive system.

Indeed, the president and his supporters at first reacted with annoyance to the use of the label and even tried to prevent the media from using it, because it was being exploited by political opponents to make the program appear to be a hopeless and dangerous adolescent fantasy. The objection was dropped as the phrase struck and the critics failed to arouse great public opposition by using it. On the contrary, the idea of a space-based shield proved to be appealing,[62] possibly because it seemed as though it would keep nuclear explosions well away from the earth, encouraging the thought that even if a war were to come to pass, it would be fought in space. Many people may also have been ready to believe that in view of what had already been achieved in space technology, future systems might well alter the character of warfare more or less as the president was suggesting. When the Fletcher panel reported in favor of a layered system that would emphasize deployments in space, and the president endorsed the report, any point in objecting to the designation of "Star Wars" was lost.

A curious omission in the president's speech is any reference to Soviet efforts in strategic defense. This omission left him open to the criticism that he was setting in motion a new chapter in the arms race which would destabilize the superpower relationship by giving the U.S. an advantage in an area which had been foreclosed by the ABM Treaty. A subsequent government publication[63] made up for this omission by emphasizing that the Soviets already had made considerably greater investments than the U.S. in defensive systems. These included more than 10,000 SAM sites to defend against the U.S. bomber force (in which more than half the megatonnage in the American arsenal is carried). The report pointed out that the Soviets were hard at work on an undeclared SDI of their own, designed to improve their existing ABM system around Moscow, perhaps to upgrade SAMs so that they might serve for BMD, to develop mobile radars, and to do research on BMD involving "new physical principles," especially beam weapons. Secretary of Defense Weinberger, in particular, stressed the need for SDI to prevent the Soviets from achieving a space-based defense first.[64]

There are several possible explanations for the omission. The drafters of the speech could not have been unaware of Soviet efforts, even though they did not give the Defense Department much time to study the speech and react to it. More likely, they were reluctant to make an issue of them because the considerable Soviet expenditures had been met by very effective U.S. offensive countermeasures, demonstrating that the effort to build defenses had so far been futile and very costly. Conceivably, the president may have deliberately omitted reference to the Soviet program in order to stress the novelty of his proposal and its benign characteristics. (He had not long before characterized the U.S.S.R. as an "evil empire." Could such an empire be expected to have taken an initiative aimed at ridding the world of the scourge of nuclear weapons?)

Buttressing the strategic rationale for research into defenses and their eventual deployment was an economic rationale. Although this did not emerge in the president's speech, which omitted any mention of the economic consequences of the initiative, it already had been stressed by High Frontier. The economic issue began to be addressed once the Fletcher panel put a price tag on the research program and critics began to estimate the cost of the program, some in the range of a trillion dollars. Upon receipt of the Fletcher Report, Under Secretary DeLauer recommended, and Secretary Weinberger endorsed, a research phase that would cost $26 billion over six years. Critics attacked this budgetary request on several grounds. They argued that it would distort research priorities in defense, because SDI would compete with other important programs aimed at modernizing nuclear and conventional forces; that it would further contribute to the "militarization of research" at universities, already under way because of the Reagan budgets' emphasis on military research and development; that it would be wasteful, because it would have to be resolved before an operational system would merit testing; that the program would divert scarce talent and industrial resources from efforts at innovation in the civilian economy at a time when the nation was suffering from "de-industrialization" and badly needed to restore its economic competitiveness; and that the real motivation for such a rapid escalation in expenditures was to create a constituency of contractors that would provide the program with enough momentum to make it difficult, if not impossible, for some future administration to scale back or curtail.[65]

In response to such criticisms, administration spokesmen made several key points. The program was based on research that had been going on for some time, so the expenditure totals should be regarded as increments. In other areas, research was no longer critical because new military technologies were in a development or deployment phase. The research budget for the SDI would amount to no more than a comparatively small fraction of total military research, and would make it possible, early in the 1990s, to decide whether to proceed with the development and deployment of a defensive system. Before then, it was purely speculative to try to determine the costs of deployment. These costs, too, would in any case be spread over many years and would therefore represent a manageable fraction of the defense budget.

In addition, advocates and administrative spokesmen also emphasized the potential importance of the SDI as a source of spillover benefits for civilian high-technology industry.[66] Advances in computers, optical sensors, materials, and other aspects of SDI research, they suggested, would stimulate a resurgence of American industry in much the same way as defense research had done in the 1950s and 1960s, and with greater indirect benefits than the more specialized mission-oriented research of the space program of the 1960s and 1970s. An office of Innovative Science and Technology was created within the SDIO to assure opportunities for broad-gauged research at universities, and other efforts were made to ensure that smaller, highly innovative companies would receive significant shares of the funding. Critics even wondered whether the SDI was in reality a covert way for a conservative administration to achieve a form of reindustrialization. The French government suspected that whatever the intention, the result would be to provide massive support to American industry that would leave Europe in the lurch, and President François Mitterrand therefore proposed the Eureka project, a mainly civilian-oriented European alternative.

While the strategic technological rationale is on the surface the main source of the SDI, and the economic rationale only a subsidiary theme, a political rationale is more likely to be its underlying source. This rationale has both ideological and pragmatic aspects.

The Reagan administration's commitment to SDI reflects its conservative ideology, which emphasizes the goal of advancing the cause of democratic capitalism against Soviet communism and imperialism. This goal is thought to require "a strong military" in virtually every conceivable respect. The ideology is unabashedly nationalistic and skeptical of the value of arms control, detente, and international organizations. Continued nuclear testing is supported on the ground that so long as the U.S. remains committed to nuclear weapons the military must be assured the weapons are effective, and researchers must be free to improve their characteristics and reliability. The ABM Treaty is disliked because it inhibits the U.S. from developing defensive weapons, making military strength a matter of developing offensive capabilities alone.

From this perspective, military R & D of all kinds is vital. Military strength can best be assured not only by providing the armed services with the hardware they need and with the ability to attract and keep the personnel to operate them, but also by encouraging the weapons laboratories to advance the state of the art. There is no room in this position for President Eisenhower's strictures against the military-industrial complex[67] or for the liberal and radical view that argues for economic conversion of military industry to civilian pursuits. A cardinal tenet of the conservative view in general, which applies especially to SDI, is the belief that America's great strength is its free enterprise economy, especially the high-technology sector of that economy. By freeing industry from the burden of high taxes, conservatives expect to stimulate enterprise. By providing government subsidies for military R & D, they aim to direct that industrial strength to the maintenance of national security. By maintaining stringent security classifications and restrictions on exports of technology to the Soviet bloc, they intend to protect the advantages of Western science and industry in the military competition with the Soviets.

The fundamental belief informing this general policy and the SDI is that the U.S. is locked in a deadly competition with the Soviet Union, and that superiority in military technology is a critical factor that will affect the outcome. The Soviets, conservatives believe, are determined to achieve military superiority, even at continued cost to the civilian sector of the economy. Unlike liberals, who tend to interpret increased Soviet military expenditures mainly as an effort to attain and maintain strategic parity with the West, conservatives hold that the Soviet aim is not parity but superiority—a superiority they would seek to exploit for political gain and if possible to achieve a military victory over the West.[68] From this point of view, detente generally serves the interest of the Soviets far more than that of the U.S. It enables them to conduct subversive proxy efforts to destabilize the West through "wars of national liberation" in the developing countries and through "low intensity warfare," including terrorism. Wherever they can, the Soviets seek to block resolution of conflicts in the developing countries, in the hope of picking up the pieces or forcing the West to protect its interests by committing resources to these conflicts. Within its own spheres of influence, in Western Europe or Asia, the Soviets remain committed to the "Brezhnev Doctrine," i.e., they aim to consolidate their hold and extend it if possible. A U.S. commitment to military improvements helps put the Soviets on the defensive, forcing them to match Western expenditures and thereby putting a strain on their resources. Such strains are apt to inhibit the Soviets from attempting to acquire new dependencies and foment difficulties for the West, especially insofar as the Soviets find that they need access to Western technology in order to keep pace, both in military and in civil terms.[69]

In this international perspective, SDI can be regarded as merely another face of the broad version of the "Reagan Doctrine"—another aspect of the effort to take the offensive in the competition with the Soviets, and to counter their efforts to intimidate the West and undermine its interests by active countermeasures.[70] The Soviets fear SDI, according to this reasoning, because they know the West can make it work and that it will degrade their efforts to achieve offensive superiority. To cope with it, they will be compelled to increase expenditures on countermeasures and defenses of their own which could put an unbearable strain on their economy. SDI, moreover, can be seen as a pragmatic response to the Soviet effort to seize the high ground of space and develop advanced weapons. It is both an effort to steal a march on the Soviets and at the same time an effort to make sure they do not do the same on the West.

This political rationale carries a great deal of appeal not only among hard-core conservatives but among a broader public attuned to the same sentiments and perceptions. Adherence to the SDI has become a litmus test not only of personal loyalty to the president but of loyalty to his future legacy. Aspirants to the Reagan mantle like Congressman Jack Kemp have made the pledge to maintain SDI a token of their commitment to the conservative agenda, along with tax reduction, and militant anticommunism. The president's excellent sense for American public opinion also led him to see the broader appeal of SDI. To those concerned by the prospect of nuclear attack but opposed to political accommodation with the U.S.S.R., it provided hope for protection. This hope, moreover, entailed the development of nonnuclear systems in the main and therefore was not open to the usual objection that increasing America's military strength would only cause an escalation in the lethality of the arms race. SDI thus responded to a widespread longing for removal of the nuclear threat without requiring progress in arms control or trust the Russians. It required only the kind of self-confidence in their own economy and their own capacities for technological achievement that Americans traditionally believed in and that had recently led to impressive achievements in space technology. Although liberals claimed it would unleash a new, no-holds-barred phase of the arms race, most Americans reacted favorably to the project, even though they also favored arms-control agreements. To young people, the SDI carried a particular appeal, because of their absorption in the romance of science-fiction-style space adventure. It must have seemed hardly as far-fetched to them as to many of their elders, who still thought of "Buck Rogers stunts" as belonging to the comic pages of their newspapers. Although critics warned that it would unleash a new, unrestrained chapter in the arms race, everywhere, even in Europe, where leaders openly expressed skepticism about its feasibility, SDI captured public enthusiasm precisely because it seemed to respond to a deep fear (ironically one that had been exacerbated by disarmament forces) with a hope that did not seem unduly provocative and warlike. A nonnuclear shield seemed a good deal less threatening than a Damoclean sword composed of 50,000 nuclear weapons. By promising to share the technology with the Soviets, and emphasizing that his intention was to achieve a defense for all mankind, and not U.S. military advantage, Reagan couched SDI in terms that gave it a very broad appeal. On SDI, as Garry Wills has observed, "[O]thers have the arguments but Reagan has the audience."[71]

As the impasse at Reykjavik late in 1986 demonstrated, both the intention behind the SDI and its implications are open to considerable confusion and disagreement. President Reagan sees it as an effort to develop a strategic alternative to reliance on the threat of offensive retaliation to deter a nuclear attack and as an effort to maintain U.S. military strength in the face of a continuing Soviet buildup. He also argues that the achievement of a comprehensive strategic defense would be of benefit to all mankind by providing protection against the damage a nuclear war would inflict should the present form of deterrence fail. Others see the project quite differently. To a legion of skeptics inside and outside the Pentagon, there is little prospect that its stated goals can be achieved in the foreseeable future. Technologically sophisticated critics see the SDI, especially that aspect of it oriented toward boost-phase and midcourse interception, as being based on nothing more substantial than wishful thinking. But even some critics concede that defenses may be advisable as a means of shoring up deterrence by retaliation—especially hard-point defenses capable of protecting retaliatory capacities against preemptive attack. Enthusiasts see the project as a way to gain superiority over the Soviets, by exploiting the U.S. qualitative advantages in high technology to seize the high ground of space, in order to prevent the Soviets from gaining a military advantage in space, to put them on the defensive, and perhaps compel them to give up other efforts to maintain superiority in strategic and conventional forces. Others see the SDI as a possible opportunity for both superpowers to recognize the need to achieve agreements on reducing arms and controlling the testing and deployment of destabilizing weapons.

It is in this last interpretation that the best hope lies for converting SDI into a much less controversial undertaking. Even some well-informed critics acknowledge that if a comprehensive defense were really feasible, it would be desirable. Although the world has survived for forty years while living under the threat of a nuclear catastrophe—some would say because of this very threat—there can be no assurance that this form of deterrence will work indefinitely. As has often been noted by specialists in strategy, the current arrangement is not perfect proof against an accidental launch of a nuclear weapon which could trigger an unintended war and could do considerable damage even if it was recognized as unintentional. More probable is the danger that a conventional war, perhaps initially involving only proxies or allies of the two superpowers, might escalate into a nuclear confrontation between them, as conventional forces are introduced with "dual capable" (conventional and nuclear) weapons and either side decides to risk using nuclear weapons on a "limited" or tactical basis. A mutually deterring defensive shield might well be preferable, even though it would remove the offensive threat with which NATO now deters Soviet conventional advantages; any extra burden that may be required to assure a conventional balance would be small compared to the catastrophic threat lurking in the potential failure of offensive deterrence, and the removal of the nuclear threat could lead to conventional arms reductions.

The desirability of defense is justification for some level of research expenditure to investigate the prospects. Even some of the severest critics of the SDI concede that research is justified in case it should lead to effective technologies and as a hedge against Soviet breakout.[72] What they object to is the projected scale of the research, which would be wasteful and could lead the Soviets to refuse either to abide by existing arms-control agreements or to enter into new ones, because the most direct response to a promising defensive program is not just to match the program but to build offenses and countermeasures capable of dealing with it—an effort which could be incompatible with maintaining adherence to the ABM Treaty and destructive of the entire effort to achieve arms control.

A commonly expressed Soviet view is that SDI is being undertaken not merely to provide the West with a strategic defense but to give the U.S. a decisive edge, by adding "space strike" weapons that could presumably be used not just for defense but for offensive purposes as well. Such systems could be used to destroy Soviet reconnaissance and communications satellites (especially those that the Soviets are developing to provide real-time targeting) and to enable the U.S. either to launch a first strike against the U.S.S.R., in the expectation of being able to absorb a second— supposedly "ragged"—strike, or to exploit its military advantage for political gain.[73] The Soviet view can claim some support from official U.S. sources. The day the president made his "Star Wars" speech, DOD officials testifying before the Senate Armed Services Committee pointed out that directed energy weapons could have offensive uses. Such weapons could "deny the use of space for collecting the distributing crisis management and targeting information while protecting our space assets that warn against a surprise attack," and they could be used to "neutralize or disrupt enemy targets" as well as to aid U.S. retaliatory strikes.[73]

The Soviets, like some domestic critics, also see the SDI as a calculated effort to force them into a competition with the U.S. economy, one that would confront them with the handicap of their relative backwardness in technology and production and impose expenditure burdens that would make it impossible to meet growing demands for an improved standard of consumption as well as improved performance in the civilian sectors of the economy. It is also possible, however, that the Soviet leadership is deliberately exaggerating the threat it sees in SDI. The technological challenge of SDI serves a useful domestic purpose in making Gorbachev's economic reforms seem critical to national security. By claiming that the U.S. aims to "militarize space," the Soviets can divert attention from their own considerable efforts to exploit space for military purposes and hope to arouse opposition to the U.S. in Europe and elsewhere. The Soviets may also be preparing the ground for a negotiating strategy in which they would offer to accommodate SDI by accepting a broad definition of the ABM Treaty provisions on research, in exchange for U.S. acceptance of Soviet activities considered violations of existing treaties or perhaps for even more valuable concessions.

The unusually political character of its initiation, however, could prove to be SDI's undoing, at least in the short run. Once the Reagan administration leaves office, SDI will become more vulnerable to attacks that have already begun to be mounted. The armed services will view large increases for SDI as competitive with other priorities. Congressmen seeking ways to control the growth of the federal budget will find SDI a prime target. Skepticism rife in the defense community over the technical prospects may be expressed more openly. And ironically, the Reagan administration's own achievements in arms control, even if limited to the signing of a treaty banning intermediate-range weapons, will leave SDI as the major obstacle in the way of broader progress toward arms reduction and better relations between East and West.

References

1. Robert S. McNamara, excerpts from a speech delivered 18 September 1967, in San Francisco before a meeting of journalists, reprinted in Abram Chayes and Jerome Wiesner, eds., *ABM: An Evaluation of the Decision for Development* (New York, 1969), 237.
2. For the role of the refugee physicists, see Richard G. Hewlett and Oscar E. Anderson, Jr., *The New World, 1939/1946, vol. I of a History of the United States Atomic Energy Commission* (University Park, PA, 1962), 14–19; Martin Sherwin, *A World Destroyed* (New York, 1975); 18–30, and J. Stefan Dupré and Sanford A. Lakoff, *Science and the Nation: Policy and Politics* (Englewood Cliffs, 1962), 91–93.
3. Truman reached his decision after the question had been well aired within the executive branch.

The General Advisory Committee to the Atomic Energy Commission had recommended against a crash program, but the commissioners themselves favored proceeding with it by 3–2. Truman asked a subcommittee of the National Security Council, composed of AEC chairman David Lilienthal, Secretary of Defense Louis Johnson, and Secretary of State Dean Acheson, to review the matter for him. The subcommittee was in favor by 2–1. Truman's decision was supported by prominent nuclear physicists, including Karl T. Compton, Edward Teller, Ernest Lawrence, John Von Neumann, and Luis Alvarez. See Herbert F. York, *The Advisors: Oppenheimer, Teller, and the Superbomb* (San Francisco, 1976).

4. Studies of defense policy-making provide support for each of these factors. The role of "bureaucratic politics," i.e., the interplay of departments and agencies in the executive, acting out of organizational interest and perspective, is often stressed. Thus, in *The Polaris System Development: Bureaucratic and Programmatic Success in Government* (Cambridge, MA, 1972), Harvey Sapolsky notes that the success of the Polaris project was a result of the skill of its proponents in bureaucratic politics: "Competitors had to be eliminated; reviewing agencies had to be outmaneuvered; congressmen, admirals, newspapermen, and academicians had to be co-opted" (244). In examining the controversy between the Army and Air Force over which service should get responsibility for intermediate-range ballistic missiles, Michael H. Armacost also notes the importance of lobbying by the services and the applicability of the pressure-group model of analysis. He points out, however, that the services found it necessary to build consensus among journalists, congressmen, analysts, in quasi-autonomous "think tanks" and "an extensive network of scientific and technical advisory committees located with the Executive branch" (256), and that this "very pluralism assured the government of a broad base of scientific and technical advice, and, superimposed upon service rivalries, this provided additional insurance that criticism of weapons programs was persistent and far from perfunctory" (256). *The Politics of Weapons Innovation: the Thor-Jupiter Controversy* (New York, 1969). Ted Greenwood, in a study of the decision to adopt the MIRV principle for warheads, argues persuasively against adoption of any single-factor analysis, noting that the decision to adopt MIRV resulted from "the complex interplay of technological opportunity, bureaucratic politics, strategic and policy preferences of senior decision-makers, and great uncertainty about Soviet activities." *Making the MIRV: A Study of Defense Decision Making* (Cambridge, MA, 1975), xv. Other studies, such as Gordon Adams, *The Politics of Defense Contracting* (New Brunswick, NJ, 1982), emphasize the role of the "iron triangle" (the federal bureaucracy, the key committees and members of Congress, and the defense contractors) in promoting military expenditures. Seymour Melman, in *Pentagon Capitalism: The Political Economy of War* (New York, 1970), 70, contends that the managers of DOD "sell weapons-improvement programs to Congress and the public." None of these factors played a significant role in the decision to begin SDI, though they may well become important when and if the research phase is succeeded by a commitment to develop and deploy an SDI system, when the stakes will be much higher.

5. McNamara, 236. Graham T. Allison has suggested, however, that "U.S. research and development has been as much self-generated as Soviet-generated." "Questions About the Arms Race: Who's Racing Whom? A Bureaucratic Perspective," in Robert L. Pfaltzgraff, Jr., ed., *Contrasting Approaches to Strategic Arms Control* (Lexington, MA, 1974), 42.

6. Herbert F. York, *Race to Oblivion: A Participant's View of the Arms Race* (New York 1970), 238–39.

7. Paul B. Stares, *The Militarization of Space: U.S. Policy, 1945–84* (Ithaca, 1985), 190–92.

8. Testimony of Robert Cooper to the Senate Armed Services Committee, *Hearing on Strategic and Theater Nuclear Forces,* 97th Congress, second session, part 7, 16 March 1982, 4845–76.

9. Reagan's solicitation of advice on SDI from "a highly selective group" intensely loyal to him has been contrasted with Eisenhower's submission of the proposal for a nuclear test ban to a broadly representative group of scientists in the President's Science Advisory Committee by G. Allen Greb in *Science Advice to Presidents: From Test Bans to the Strategic Defense Initiative* (San Diego, Research Paper No. 3, 1987), 15.

10. For a detailed account of how this decision was reached, see Samuel P. Huntington, *The Common Defense* (New York, 1961), 326–41.

11. Interview in *Newsweek*, 18 March 1985, quoted in *Star Wars Quotes*, compiled by the Arms Control Association (Washington, D.C., 1986), 26.

12. Donald Bruce Johnson, compiler, *National Party Platforms of 1980* (Urbana, 1982), 207.

13. Garry Wills, *Reagan's America: Innocents at Home* (Garden City, NY, 1987), 361.

14. Martin Anderson, *Revolution* (San Diego, 1988), 83.

15. Text of interview in Robert Scheer, *With Enough Shovels: Reagan, Bush, and Nuclear War* (New York, 1982), 232–33.

16. Greve.

17. The plan was publicized in July 1983 by *Aviation Week & Space Technology*. Stares, 219.

18. Greve.

19. Remarks to reporters, *Weekly Compilation of Presidential Documents* (Washington, D.C., 4 April 1983), 19.3, 453.

20. Daniel O. Graham, "Towards a New U.S. Strategy: Bold Strokes Rather Than Increments," *Strategic Review* 9.2 (Spring 1981), 9–16.

21. Lt. Gen. Daniel O. Graham, *High Frontier, A New National Strategy* (Washington, D.C., 1982), 18.

22. Ibid., 9, 20.

23. Testimony of Robert Cooper, 4635.

24. "DOD;s Space-Based Laser Program—Potential, Progress, and Problems," Report by the Comptroller General of the United States (Washington, D.C., 26 February 1982), iii–iv.

25. In 1982 DOD established a Space Laser Program as recommended by DARPA, in cooperation with the Air Force and the Army. The plan called for the expenditure of $800 million over the period from FY 1982 through FY 1988, under the supervision of the Office of the Assistant Secretary for Directed Energy Weapons.

26. Gregg Herken, *Cardinal Choices: The President's Science Advisers from Roosevelt to Reagan*, draft of chap. 6, p. 21. A substantial part of this chapter has been published as "The Earthly Origins of 'Star Wars,'" in *The Bulletin of the Atomic Scientists (October 1987)*, 20–28.

27. Pete V. Domenici, "Towards a Decision on Ballistic Missile Defense," *Strategic Review* 10.1 (Winter 1982), 22–27.

28. William J. Broad, *Star Warriors* (New York, 1985), 75–79.

29. Edward Teller, "SDI: The Last Best Hope," *Insight* (20 October 1985), 75–79.

30. Broad, 122. "In all," according to Broad, "Teller met with the President four times over the course of little more than a year." Teller, in a letter to the authors (21 September 1987) claims to have had little direct influence on the president's decision: "Before the President's announcement of SDI, I had two very brief meetings with the President. I expressed no more than my general support and good hopes." In the September meeting, he recalls, "defense was mentioned but no subject like the X-ray laser was explicitly discussed." With respect to the X-ray laser, Teller's recollection does not jibe with Keyworth's. See Herken, 23.

31. Quoted by Greve.

32. Quoted by Broad, 73.

33. Edward Teller with Allen Brown, *The Legacy of Hiroshima* (Garden City, NY, 128–29.

34. In an address to the Faculty Seminar on International Security, UCSD, 12 December 1983.

35. Greve.

36. Statement by the Assistant for Directed Energy Weapons to the Under Secretary of Defense for Research and Engineering before the Subcommittee on Strategic and Theater Nuclear Forces of the Committee on Armed Services, U.S. Senate, 98th Congress, First Session, 23 March 1983.

37. *Arms Control Reporter*, 10 November 1983.

38. Interview in *Government Executive* (July-August 1983), quoted in *Star Wars Quotes*, 34.

39. Greve.

40. Ibid.

41. Herken, chap. 6, pp. 46–47.

42. John Bardeen, letter in *Arms Control Today* (July-August 1986), 2.

43. Greve. Shultz is reported to have berated Keyworth for encouraging the president to believe in his vision of a perfect defense: "You're a lunatic," he is said to have told Keyworth. Hendrick Smith,

The Power Game: How Washington Works (New York, 1988), 612–614.

44. Ibid.

45. Ibid.

46. Testimony of Robert McFarlane, Defense Policy Subcommittee, U.S. Congress House Armed Services Committee, 17 May 1988, typescript, 167–68.

47. Greve.

48. Interview, *US News and World Report*, 18 November 1985, 30.

49. Text in the *New York Times*, 24 March 1983.

50. The CIA view was presented in the testimony by Robert M. Gates and Lawrence K. Gershwin before a joint session of the Subcommittee on Strategic and Theater Nuclear Forces of the Senate Armed Services Committee and the Defense Subcommittee of the Senate Appropriations Committee, 26 June 1985. The Joint Chiefs' view is reported in Strobe Talbott, *Dead Gambits: The Reagan Administration and the Stalemate in Nuclear Arms Control* (New York, 1984), 224.

51. Helmut Sonnenfeldt, address at Lawrence Livermore National Laboratory, 29 November 1986.

52. Thomas C. Schelling, "What Went Wrong with Arms Control," Foreign Affairs (Winter 1985-86), 224.1

53. See Michael Novak, "Moral Clarity in the Nuclear Age," *National Review*, 1 April 1983, 354–62, and "The Bishops Speak Out," National Review 10 June 1983, 674–81.

54. *New York Times*, 24 March 1983.

55. Herken, chap. 6, p. 49.

56. *New York Times*, 24 March.

57. Ibid.

58. Ibid.

59. Excerpt from an interview with Andropov in *Pravda*, 27 March 1983, quoted in Sidney D. Drell, Philip J. Farley, and David Holloway, *The Reagan Strategic Defense Initiative: A Technical, Political, and Arms Control Assessment* (Stanford, 1984), Appendix B, 105.

60. McGeorge Bundy, George F. Kennan, Robert S. McNamara, and Gerard Smith, "The President's Choice: Star Wars or Arms Control," *Foreign Affairs* (Winter 1984–85), 63.2, 270–72.

61. Interview with six journalists, *Weekly Compilation of Presidential Documents* (Washington, D.C., 4 April 1983), 19.13, 471.

62. A 1985 Sindlinger Poll found that 85 percent of the U.S. public favored development of a missile defense "even if it can't protect everyone." Jeffrey Hart, "A Surprising Poll on Star Wars," *Washington Times*, 9 August 1985. A Gallup Poll in November 1985 found 61 percent in favor of the U.S. proceeding with SDI. *Christian Science Monitor*, 21 November 1985.

63. *Soviet Strategic Defense Program* (Washington, D.C., October 1985).

64. On 7 December 1983, Secretary Weinberger was quoted in *The Wall Street Journal* as saying: "I can't imagine a more destabilizing factor for the world than if the Soviets should acquire a thoroughly reliable defense against these missiles before we did." Cited in *Star Wars Quotes*, 52.

65. See William D. Hartung, Robert W. DeGrasse, Jr., Rosy Nimroody, and Stephen Dagget, with Jeb Brugman, *The Strategic Defense Initiative: Costs, Contractors, and Consequences* (New York, 1986); John P. Holdren and F. Bailey Green, "Military Spending, the SDI, the Government Support of Research and Development: Effects on the Economy and the Health of American Science," *F.A.S. Public Interest Report*, 39.7 (Washington, D.C., September 1986); and Daniel S. Greenberg, "Civilian Research Spinoffs from SDI Are a Delusion," *Los Angeles Times*, 9 September 1986.

66. The potential economic benefits were stressed in Graham, *High Frontier*, 89–98, and have been outlined recently in *Report to Congress on the Strategic Defense Initiative* (Washington, D.C., April 1987), viii, 2–5.

67. "You know, we only have a military-industrial complex until a time of danger, and then it becomes the arsenal of democracy. Spending for defense is investing in things that are priceless—peace and freedom." President Ronald Reagan, State of the Union Address, 6 February 1985, *Weekly Compilation of Presidential Documents*, (Washington, D.C., 11 February 1985), 21.6, 143.

68. For one presentation of this view, see Norman Podhoretz, *The Present Danger* (New York,

1980), 56–57.

69. Ibid.

70. For an exposition of the "Reagan Doctrine," see Jeane J. Kirkpatrick, "The Reagan Doctrine and U.S. Foreign Policy" (Washington, D.C.. 1983), and "Implementing the Reagan Doctrine," National Security Record No. 82 (Washington, D.C., August 1985). This doctrine is examined critically in Stephen S. Rosenfeld, "The Guns of July," *Foreign Affairs* 64.4 (Spring 1986), 698–714.

71. Gary Wills, *Reagan's America: Innocents At Home* (Garden City, NY, 1987), 360. The broad appeal of SDI is noted by Kevin Philips, "Defense Beyond Thin Air: Space Holds the Audience," *Los Angeles Times*, 10 March 1985.

72. Thus, Sidney S. Drell has called for "a prudent, deliberate and high quality research program within ABM Treaty limits" at a level of roughly $2 billion per year. "Prudence and the 'Star Wars' Effort: Research Within the Bounds of ABM Treaty Can Aid Safer World, *Los Angeles Times*, 10 March 1985.

73. See especially Yevgeni Velikhov, Roald Sagdeyev, and Andrei Kokoshin, eds., *Weaponry in Space: The Dilemma of Security* (Moscow, 1986), chap. 4, pp. 69–77, and chap. 7, pp. 106–27.

74. "The Department of Defense Directed Energy Program and Its Relevance to Strategic Defense," Statement by the Assistant for Directed Energy Weapons to the Under Secretary of Defense for Research and Engineering before the Subcommittee on Strategic and Theater Nuclear Forces of the Committee on Armed Services, United States Senate, 98th Congress, First Session (23 March 1983), 3.

WINDS OF
CHANGE

Minimum Deterrence

I will make some "remarks" about minimum deterrence. I choose this particular word, because I will present neither a complete analysis nor a sermon. To me minimum deterrence is minimal in two different senses: one in terms of its goals and the other in terms of its means.

In terms of goals, the purpose is to deter the use of nuclear weapons by someone else and not something broader than that. And in terms of means, minimum deterrence involves very small numbers. To give my conclusions away at the beginning and then explain them in more detail as I go on, I am thinking in terms of about 100 nuclear weapons, not the 600 that the French or the British seem to think are needed for minimum deterrence or the several thousand that are suggested for the American forces later at the end of this century, but something in the neighborhood of 100. Minimum deterrence necessarily involves plans for the use of these weapons against the potential enemy's most precious targets, which is another one of the euphemisms that we always use when we talk about nuclear weapons. A clearer translation of this notion of most precious targets is cities, which of course raises major moral issues that I will also discuss later.

As I see it, the notion of minimum deterrence is entirely consistent with the idea of no first-use. In fact, it is almost the same idea. Whether it is consistent with extended deterrence depends on what is meant by extended deterrence. The notion of minimum deterrence is obviously not consistent with the idea of using nuclear weapons in the event someone else commits a major non-nuclear act, but it is not necessarily inconsistent with the notion that one might provide some kind of a nuclear umbrella to a particular ally or group of allies—against a nuclear attack and only against a nuclear attack against them.

Like mutual-assured destruction, minimum deterrence is not an objective that is intrinsically desirable, it is simply the best of what is practically obtainable from a bad set of possibilities. Supporting it is a recognition of the facts that nuclear weapons, and the terrible threats that inhere in them, simply cannot be eliminated from the world. To be clear with regard to my view about why they cannot be eliminated, I am thinking in terms of technology. A world that can produce modern aircraft and solid-state thin-screen television can readily produce or reproduce nuclear weapons, even if they were all somehow destroyed at one particular moment.

In these remarks, I will try and give some idea of the derivation of the number that I presented. I will comment further on the moral issue that the notion of deliberately targeting the most precious targets raises, and I will also make some remarks about stability. These last two matters, in particular, are often raised as objections to strategies of minimum deterrence. I want to acknowledge the fact that I am aware of those objections and have at least the initial elements of an answer to them.

Number of Nuclear Weapons

With regard to the number, I arrived at it by first working up from zero, then working down from where we are, and discovering something in between that fits my particular set of notions. With regard to working up from zero, I have long regarded the remarks that McGeorge Bundy made, shortly after he left his position in the mid-1960s as National Security Advisor to Presidents Kennedy and Johnson, as cutting to the core of the situation. In sum, he said that, in the real world of real political leaders, any policy, which in advance was known to bring a single hydrogen weapon on a city of one's own country, would be regarded in advance as a catastrophic blunder. Ten weapons on one's own country would be a disaster beyond history, and then he ended by saying that 100 is unthinkable. So the number that is necessary to deter at least the people that I have some direct understanding of (Western and American leaders) is somewhere in the range of 1, 10, or 100. I think that it is closer to 1 than it is to 100, but of course there has be some reasonable assurance of the likelihood that nuclear retaliation really would happen for that deterrence to be effective.

I am not sure about other leaders. Names of supposedly irresponsible third world leaders are often thrown out, and the suggestion is made that the people who took so many tens of millions of casualties in World War II cannot be deterred merely by the threat of the loss of a single city, or that there will always be crazy people out there. And there may very well be irrational people with nuclear weapons, but I do not think they practice the kind of calculus that is relevant to the notion that even 5000 deter. Thus, as far as rational people approaching the question of nuclear weapons falling on themselves, somewhere between 1, 10, and 100 is the answer to what it takes to deter, and for irrational leaders there may be no answer at all.

Coming from the top down, one criticism of reductions which people who are fairly passionate about these questions sometimes make, is that it does not make any difference. Drop the number from 12,000 to 6000, and the effect of 6000 on civilization and mankind is nearly the same as the effect of 12,000, so it does not make any difference. Or drop from 6000 to 3000 and it does not make much difference either, because 3000 weapons still contain such an enormous capability for death and destruction. I admit there is something to that notion although it clearly depends on precisely how they are targeted. But there is a point at which reductions begin to make a real difference, and that point is where the number of weapons has been cut to the point that no matter how the leaders decided to use them, they could not produce something substantially worse than World War II. I do not know exactly where that point is, but I estimate that is somewhere in the neighborhood of 100. I am sure it is not in the neighborhood of 1000. It may in truth be less than 100, but it is somewhere in that neighborhood.

I simply cannot be more precise than this, but I have given you my two ways of approaching this number, coming from the bottom up and trying to think of what it would really take to deter an American President or someone else who thinks rationally, and coming down from the top and trying to arrive at a number where the reduction really would make a difference, and where no matter what the later decisions might be, the result would be no worse than, as I said, World War II.

Moral Issues

Turning to the moral issues, it is commonly said that deliberately targeting cities is simply unacceptable and that we must have some kind of strategy based on destroying military targets. My particular answer to this problem is that the other strategies, the strategies that involve so-called military targets or counterforce strategies, always necessarily involve a much larger number of nuclear weapons. By the time one looks at what the military targets (usually called military/industrial targets) are in any real plan, one finds, for instance, that they include all the naval bases, which include most American big cities on the coasts. One finds that they also include the major centers of command, communication, transportation, etc., and that striking these military/industrial targets in a way designed to remove the enemy's military capability leads to a number of potential civilian casualties that is at least as large as the number that would be inherent in a deliberate attack involving a smaller number of weapons against the cities themselves. The main difference is that, in the case of an attack against the military targets, people are collateral damage; and, in the minimum deterrence case, they are deliberate damage. I do not see an important moral distinction between the two cases. I do agree with the view of the Catholic bishops that the use of nuclear weapons against the large civilian populations and the threat of their use are both wrong. But the situation, as I see it, is that we are currently faced with a number of choices, all of which are bad, which is why I said at the beginning that I do not regard minimum deterrence as a desirable objective, but simply the best that is within reach.

Stability

My last point concerns stability. It is often said that small numbers are intrinsically less stable than large numbers, and there are various reasons given for that claim. For example, if we had an agreement covering what the allowed number would be, then cheating by 100 at a level of 10,000 would be one thing, and cheating by 100 at a level of 100 would be quite different. This is true when discussing force-against-force, in the standard counterforce situation, but if we restrict our thinking to discussing minimum deterrence and countervalue targeting, then I believe it is possible to solve this particular problem by putting special technical emphasis on force survivability. When one gets down to numbers this small, the survivability question really becomes quite different from the one we have become familiar with, and I think that it becomes amenable to more direct and more certain answers. To give just two examples: one might deploy these 100 as single weapons on single missiles with each missile deployed on a separate ship; or one might deploy them in a larger number of holes (i.e. missile silos) located two miles or more apart out in Nevada on military reservations, with the total number of holes equal to twice the number of nuclear weapons, not just those in the hands of a potential enemy, but twice the number of nuclear weapons that all other countries have together. If one is down to something like 100 weapons per country, then multiplying all nuclear weapons in the world by two still leads to a number of holes that is feasible to put somewhere like that. I am not trying to work out the technical details here, but I do think that the notion of survivability becomes so different at very small numbers than it is at large numbers, that the simple idea that it is necessarily less stable because cheating becomes too significant, does not apply.

Summary

I do not regard these remarks to be a comprehensive analysis of the situation, but a discussion of the future of nuclear weapons ought to have somebody paying some attention to this particular goal, which is in fact held by a great many people whom I regard as serious. Of course, the possibility of moving so far is, to be realistic, very poor at the present time. It is clear that numbers more like those Mike May and others talk about (5000 or so) are much more realistic, much more possible in the near term, and I agree with the kind of analysis that leads to those numbers for the near term. The place that I disagree is when they say that this is a long-term asymptote. I think of this (5000) solely as an intermediate number, which may in fact last for a number of years, but I do not think that it is the proper asymptotic number for the number of nuclear weapons.

Nuclear Arms Race:
Past, Present, and Future

The roots of the nuclear arms race lie both in the great European physics of the fifty years leading up to 1939 and in the terrible political events that convulsed the world continuously 1914 to 1945. The end of that awful period was punctuated by the first explosion of three atomic bombs; one in New Mexico and two over Japan. Two bombs were made of plutonium, the substance produced as the direct result of the events we remember here today. The other, the Hiroshima Bomb, was produced by entirely different means. Even if the events that took place in Chicago on December 2, 1942, had failed or proved to be impossible, we would still not have entirely escaped the grave dangers we lived with during the last several decades.

After World War II ended, the huge American effort that produced those bombs, the Manhattan Project, was largely, but not entirely, demobilized. The Soviet program, which had begun at the same as the American program but on a very much smaller scale, was revitalized and greatly expanded, and soon was able to reproduce the American result in about the same length of time. Taken together, these counterposed activities could be called a nuclear arms race, although they were on a modest scale compared with what was to come.

In 1949–1950, three closely spaced events brought about a revolutionary change in the world situation and initiated the great nuclear arms race that would hold the world in its thrall for the next forty years. These events were the successful detonation of an atomic bomb by the Soviets, the formal establishment of the Sino-Soviet Bloc, and the Korean War.

The first of these events, the explosion of a Soviet bomb, was not a surprise to most of the insiders. but it was to the public as a whole, and it produced a singularly bad shock in Washington and other Western capitals. It was widely seen as a challenge that must be met and being nuclear in nature, a nuclear response seemed appropriate. Proposals were, therefore, elaborated for substantially expanding the American program in all of its dimensions and for pursuing the development of a still bigger bomb, known variously as the hydrogen bomb, the thermonuclear bomb or, simply the "Super." Expansion of the program was generally endorsed by all those involved; however, a high priority pursuit of the Super was, as is well known, opposed by David Lilienthal, then chairman of the Atomic Energy Commission, and by some American nuclear scientists, most notably Robert Oppenheimer and his colleagues on the General Advisory Committee to the US Atomic Energy Commission.

The second of these events involved the proclamation of the Peoples Republic of China just one month later by Mao Zedong in Tian An Men Square and the subsequent prolonged meeting of Mao with Stalin in Moscow the following January and February. We still don't know much about the details of that meeting, but when it ended various proclamations and reports about it appeared in *Pravda*. Among other things they told us:

> This (new alliance) cannot help but alter the world situation and strike a solid blow against the aggressive plans and policies carried on throughout the world by the imperialist bloc, led by American imperialism. In other words, the Sino-Soviet treaty changes the balance of forces.

But despite the challenge apparent in this new situation, the Defense budget submitted to the Congress by Harry Truman that spring was the lowest since before the war.

The third, and decisive, event was the invasion of South Korea by North Korea in June 1950, only some nine months after the Soviet atomic bomb and only three months after the declaration of the Sino-Soviet Alliance. To anyone even half aware of what was happening, this seemed to be the confirmation of the Mao-Stalin promise to "strike a solid blow against...American imperialism" and to exploit the "changes (in) the balance of forces."

The immediate response was the Korean War.

The long term response was a complete overhaul of our national security policy and the initiation of a full scale arms race to build the means we believed necessary for maintaining peace, and assuring our survival, in this new world.

Even before the Korean War was over, we in the United States took a "New Look" at our strategy and concluded first and foremost that there must be no more "Koreas," that never again should we allow the aggressor to pick the place and the weapons that best suited him. Next time, we would respond to such an action in a place favorable to us and with whatever weapons gave us the advantage. In brief, we would strike the source of the problem and we would employ high technology, our strongest suit, rather than massive manpower, which we saw as theirs. The name for this new approach was "Massive Retaliation." The remobilization of science, in a manner similar to what we had accomplished during World War II, but on an even larger scale, was to be the means for carrying out this new strategic policy.

The rest, as they say, is history. Within ten years we designed, built and deployed 32,000 nuclear weapons, nearly all of them involving thermonuclear processes, and we constructed and maintained ready more than 2,000 "strategic nuclear delivery vehicles"—long range aircraft, intercontinental rockets and submarine launched ballistic missiles—capable of delivering the bombs and warheads to targets anywhere in the world, in most cases in less than half an hour. We also deployed more than 7,000 so-called tactical and battlefield nuclear weapons in Europe, with additional weapons in Korea and at sea. All this was in keeping with our policy of "extended deterrence," as it was called by then. From that time (the early 1960s) on, all further changes in both policy and hardware were strictly evolutionary, being either in reaction to further marginal changes in the behavior of the USSR or deriving from the steady march of technological change.

Something similar was happening in the USSR. They, too, built about the same number of weapons and delivery vehicles and deployed tactical weapons at sea and in forward, out of country, locations. However, as a consequence of their weaker technological base, it took them about ten years longer to do so. In part, their program was a response to ours, and in part it had its own internal logic. We still cannot sort out these factors in a convincing way. But in addition to the events of the early 1950s recounted above, their plans were uniquely affected by the Cuban missile crisis, in which they felt they had been humiliated by us. The massive buildup during the Brezhnev era surely is due at least in part to this last factor.

Fortunately for the whole world, the top national leadership or both sides (one might well say "all sides") recognized that nuclear weapons were simultaneously and intrinsically both a solution and a problem. They seemed to be the best, and sometimes the only, available solution to certain immediate, severe and otherwise intractable security problems. At the same time it also was clear that "maintaining peace by threatening mutual suicide," to put it bluntly but realistically, was intrinsically dangerous and could not possibly be a long term solution to anything. These two contrary aspects of nuclear weapons led to contradictions that simply could not be fully resolved. But happily for the rest of us, our leaders found ways to live with them even so.

Many, perhaps most, members of the public as well as many top level advisors saw things differently. They commonly saw nuclear weapons as either a solution or a problem but seldom as both. Some believed they were a perfectly adequate solution to our problems, and that we only undermined our security by overemphasizing their dangers or by trying to negotiate their control with an implacable enemy. Others advocated getting rid of them before even beginning to resolve the problems that gave rise to them in the first place. The leaders, however, from Truman to Reagan and from Khrushchev to Gorbachev were fully aware that "a nuclear war cannot be won and must not be fought" and behaved accordingly. Thus, on the one hand, all American presidents, from Truman and Eisenhower to Reagan, made certain that there were enough weapons and that they were of adequate quality, their common goal being a level of technological superiority adequate to deter and make up for the larger numbers of men and conventional weapons on other side. And, at the same time, they all pursued arms control objectives: the Baruch Plan, Atoms for Peace, "Open Skies," limited and comprehensive nuclear test bans, SALT I and II, the Vladivostok accords, the Non-proliferation Treaty, the elimination of intermediate range weapons in Europe, and others.

Everything I know about our own leaders leads me to believe that all of them were sincere in their approach to this matter. And everything I have been able to learn about the top Soviet leaders leads me to a similar conclusion.

Why, then, was so little accomplished in the forty years of negotiations prior to 1985?

The basic reason is that each side approached Arms Control negotiations in a spirit characterized by deeply negative convictions about the other side's near term behavior and long term intentions.

In America, and in the west generally, these convictions had deep roots, going all the way back to the establishment of the Soviet state and the savage political repression that soon followed. During World War II the fight against a common enemy softened these convictions for a while, but they were quickly reestablished afterwards. High on the list of particular causes were the events surrounding the Yalta agreement and its subsequent implementation. Broadly speaking, we intended the agreement to be the guide to the reestablishment of peace and democracy in Europe, including specifically the rebirth of the formerly independent nations of the central region. Instead, what came into being was a set of satellite states under both the political and military control of Moscow. This situation was never fully accepted by the West, although some states sometimes seemed to acquiesce to a limited extent. Nor had America in particular ever accepted the annexation of the Baltic nations, and the situation in central Europe only made that still more unacceptable. Things got worse as time went on. The sequence of Soviet armed interventions in the DDR, Hungary and Czechoslovakia made their long term intentions with respect to all of these states even more obvious. The promulgation of the "Brezhnev Doctrine" —the name attached to the political justification for these interventions after the move into Czechoslovakia—was interpreted in the West as really meaning "what's ours is ours and what's yours is up for grabs." Military and political adventures outside Europe—in Cuba, Angola, Yemen, Egypt and elsewhere— made things still worse.

Added to these political and military events were the nearly universal attitudes towards Stalin and Stalinization. Stalin was widely seen as one of history's worst tyrants. When Khrushchev denounced some of his acts at the Twentieth Party Congress, there was a tentative sigh of relief, but when he failed to follow up with anything except eliminating the most extreme forms of repression, attitudes returned almost to where they had been before. In sum, the notion of the Soviet Union as an "Evil Empire" was common long before President Reagan spoke those words. Most American and NATO leaders had long held similar views, but in the interest of not making relations worse than they already were, as well as in keeping with general diplomatic norms, they refrained from saying so in public.

These deeply held convictions poisoned relations between the two blocs across the board. Arms Control negotiations faced an additional set of specific problems. The Soviet Union was by far the world's largest country and the greatest part of its territory was off limits to all foreigners and to most of its own population as well. Secrecy was the rule in all aspects of political life, and the state security forces seemed able to maintain absolute control over what people said (and even thought) and what the media reported. The possibilities for cheating on any agreement seemed obvious. As time went on, technical means for penetrating this vast curtain of secrecy were developed, but confidence in their efficacy was never very deep among either the public or the political elites.

For all these reasons, it was extremely difficult for American presidents to make arms control proposals that went much beyond placing caps or other limitations on arms at levels which we didn't much want to surpass anyway.

And what about inside the USSR? We still don't know all the facts, but we have learned enough, especially in recent years, to make some good guesses. For one thing, as we now understand it, they were afraid of our intentions also. For years Soviets officials, including military leaders, in various fora repeatedly told us that they feared we would some day attack them unless they maintained high levels of conventional forces and an adequate nuclear deterrent. However, our leaders simply "knew" that we had no such intention, and so they concluded that Soviet statements to the contrary were simply one more instance of propaganda. Soviets in turn would sometimes elaborate on their claims by reciting their version of the facts surrounding the establishment of the Soviet state in the first place, but most of our people remained unconvinced. Finally, in the late 1980s, after a number of mutual visits by the military leaders of the two blocs, visits that often included serious personal and professional exchanges, many of our military came to understand that yes, surprising as it may seem, they really had a basis for concluding the worst about our intentions towards them.

In at least one important regard the two sides approached the question of arms control quite differently. On our side, civilians of all kinds and stripes were involved in strategic thinking, in developing ideas about military policies, and in formulating the details of our negotiating proposals. These included civilian advisers in the White House and at almost all levels in the military departments, the experts in numerous "think tanks," the civilian staffs in the Pentagon and the State Department, scholars in many universities, pundits on the editorial staffs of the media, members of Congress and their assistants, and so on and on.

In the Soviet Union the situation was totally different. At least until 1982–83, strategic thinking and the elaboration of military policies were the exclusive prerogative of the uniformed military. Similarly, virtually all decisions concerning the development and deployment of military equipment (nuclear as well as conventional) were exclusively in the hands of the military industrial complex as a whole. Even members of the Central Committee of the Communist Party largely stayed out of policy making in this area, and so did everyone else. At the formal negotiating sessions when, for example, our civilians tried to talk with theirs about the details of the strategic balance, military members of the Soviet delegation objected that these matters were secret and not the business of the civilian members. Of course, Soviet diplomats dealing specifically with these matters did discuss them among themselves and with their western counterparts, and Soviet scholars at a few select centers like the Institute of USA and Canada Studies and IMEMO (the Institute for World Economics and International Relations) discussed them among themselves and with foreign visitors. In addition, those Soviet intellectuals privileged to attend such unofficial international fora as Academy to Academy consultations on arms control and the Pugwash and Dartmouth Conferences did the same. Until the early 1980s, however, such persons played no role in determining Soviet military policy and only selected diplomats played even limited roles in determining arms control policy. (When I served as an official US negotiator at the end of the 1970s I was surprised to discover that there was virtually no contact between the Soviets I was now dealing with and those I had met earlier in discussions about arms control at the institutes and in the conferences mentioned above.) Fortunately, when in the late 1980s it finally became possible for civilians to participate seriously in strategic and other military matters, many of the people who had participated in these relatively rare and usually informal discussions (which nearly always revolved around Western data and Western ideas) were ready, eager, and able to do so in an official capacity. Indeed, in retrospect I now believe that the preparation of cadres of informed civilians and the introduction of at least a few key Western ideas into Soviet thinking was probably the most valuable outcome of those earlier interactions.

It is now completely clear, however, that until the "New Thinking" was forced upon them in the late 1980s, the Soviet military took an even more cautious view of such negotiations than ours did, and Soviet proposals were also largely limited to placing caps on deployments well above the levels that existed at the time. Evidently, more far reaching proposals, such as those made in the late 1950s and early 1960s for "General and Complete Disarmament," were only made when it was certain they would not be accepted.

The Comprehensive Test Ban negotiations may be an exception to this general rule. Soviet informants say that Gromyko and other high ranking civilians believed such a test ban was essential for inhibiting the proliferation of nuclear weapons to other states, and they did evidently manage to persuade the military to accept the possibility of at least a three year moratorium in the late 1970s.

In addition to all these general problems, certain specific events also interfered with the arms control process. Most important, the Soviet invasion of Czechoslovakia delayed the SALT negotiations at least two years, and the invasion of Afghanistan made it impossible for President Carter to accomplish any of his relatively broad arms control agenda.

Present Trends

In the early 1980s the winds of change began to stir in the Soviet Union. In the early 1990s they reached hurricane force. The result was one of history's most dramatic and most unexpected revolutions. The primary causes of the revolution were economic and political in nature, and so were its principle internal effects. In sum, Marxism-Leninism, with its command economy and so-called "democratic centralism," had simply failed to provide a decent and fruitful way of life to those ruled by it. The agents of revolution were Andropov who came to power in 1982, then Gorbachev who eventually succeeded him in 1985, and finally Yeltsin. The first sought only to reform the old system, the third finally overturned it.

But in addition to great changes in economic and political forms and behaviors, the revolution brought with it major changes in military strategy and in both internal and external security policies, and these changes in turn reflected back onto the main revolution itself.

One event and one non-event in the military area marked the earliest phase of this revolution. These were the stalemate in Afghanistan, and the lack of any military intervention in Poland during the Solidarity crises.

When Andropov took over from Brezhnev the "Period of Stagnation" came to an end and the possibility of rethinking all sorts of matters for the first time in almost twenty years arose. By a fortunate coincidence, at just that same time the stalemate in Afghanistan was becoming evident to the Soviet leadership, military as well as civilian. This stalemate undermined the confidence of the civilian leadership in the military's exclusive handling of its affairs, and helped to make it possible in this new context for high level civilians to question military wisdom and to insert themselves into strategic thinking and policy planning. There were other more subtle changes. Information coming from former Soviet civilian officials and scholars now reveals that it was also at about this same time, and for the same reasons, that such people first had a serious opportunity to discuss military strategy, weapons developments and arms control policy with high level officials and even to have some real influence in these matters.

The critical non-event, the failure of the Soviets to intervene in Poland, was, like most non-events, much harder for foreigners to observe and it therefore took longer for it to influence Western policy makers. In fact, this change in Soviet policy—the abandonment of the so-called Brezhnev Doctrine—which it portended was not made final, overt and official until Gorbachev did so a few years later at the European Parliament in the summer of 1989.

In 1985, when Mikhail Gorbachev became General Secretary, the pace of change accelerated. "New Thinking" about strategic issues became official policy, and new fresh people were allowed to participate in it. Prime Minister Margaret Thatcher said she could "do business with this man." Reagan and Gorbachev met at Reykjavik and talked seriously about truly substantial steps in arms reductions. These didn't happen right away, but new standards had been set. Soon after, the INF (Intermediate Nuclear Forces) treaty was agreed, and the START negotiations moved ahead easily. In 1989 Gorbachev, in a speech before the European Congress, declared the Brezhnev Doctrine void, and by the end of that year a "velvet revolution" had freed all the satellites from Soviet political and military domination. In effect, the Warsaw Pact was dissolved. By the summer of 1991 the Communists' monopoly on power in Moscow had also been broken and by the end of that year the Soviet Union ceased to exist.

Since that time much talk has been devoted to the topic of "Who won the Cold War." This is, of course, only natural. Even so, I think such talk is, in the main, misleading. The important fact is that we, all of us everywhere in the world, have survived a very dangerous period. Frustration, fear, wrong or garbled information about someone else's actions or near-term intentions, could at a number of critical junctures have caused one or another of the world's leaders to take an action which could have led to general or nuclear war. But it didn't happen. Out of an unknowable combination of "fear of the bomb" and general good sense, none of them ever took such a fateful step. Therefore, instead of trying to determine which of the world's leaders won or lost the Cold War, we should be praising them for maintaining the longest, the most dangerous and the most necessary peace of modern times.

One of the many results of the mighty changes described above was that the arms race went into reverse. Unilateral actions far outpaced negotiations. Indeed, by the time the U.S. Senate ratified the Start Treaty in September 1992, a treaty which called for reducing the number of strategic weapons to about 8,000, both countries were already carrying out plans for reducing to 3,500. We are also currently well into the process of eliminating most nonstrategic varieties of nuclear weapons, including battlefield systems and weapons for war at sea. And, in addition to these reductions in numbers and types, both sides have backed away from the alert postures they formerly maintained. Remarkably, the lack of adequate facilities in both countries for dismantling weapons has become the pacing factor in implementing reductions.

Various defense oriented study groups, in and out of government, are considering the possibility of still deeper cuts in the number of deployed weapons, and numbers as low as a thousand—or even less— are no longer thought of as ridiculous. Even before the collapse of the Soviet Union, environmental and safety concerns had forced the United States to shut down its plants for producing tritium and for machining plutonium parts for weapons. Now the plans for creating new facilities to do these jobs have been scaled back, delayed or abandoned altogether.

In the mid 1980s America carried out an average of twenty nuclear tests per year. Recently, this number had dropped to about eight per year. Now new legislation calls for a nine month moratorium on testing, to be followed by a period of three and a half years in which fifteen tests would be allowed solely for the purpose of assuring of weapons' safety and reliability. After that, a permanent moratorium would begin if "no other state is testing weapons" at that time. Even before the Congress forced the President Bush's hand in this matter, he had cancelled all further tests for any purposes other than assuring safety and reliability. President Clinton has continued these policies. The weapons laboratories at Livermore, Los Alamos and Albuquerque remain at their former size, as do their Russian counterparts at Arzamas and Chelyabinsk, but all of them are experiencing major reductions in the funding for their nuclear weapons programs and diligently looking for new kinds of R & D programs to replace the old ones.

Perhaps even more important than these concrete actions, Western defense intellectuals, including those in positions of authority, are coming to the view that the "problem" aspect of nuclear weapons is fast becoming much more important than the "solution" aspect. That is, a new nuclear consensus is forming around a pair of ideas: first, that when all the remnants of the old cold war have been eliminated then the United States will have no foreseeable security problems that cannot best be solved by nonnuclear means, and second, that nuclear weapons in the hands of renegade states will eventually present us with some very serious problems for which we currently have no simple solutions.

Parallel events are taking place in Russia. It, too, is carrying out similar reductions in numbers, types and deployments. President Yeltsin has recently announced an extension of the Russian moratorium on nuclear weapons tests to mid 1993. In addition, the Russians have the unique problem of bringing home all weapons from other former Soviet republics. Apparently, this has already been accomplished for all so-called tactical systems and arrangements are being developed for doing the same for strategic systems. (The situation in the Ukraine is unclear at the time of this writing.) Nuclear weapons experts and diplomats from the Russia and the United States meet frequently (both officially and unofficially) to discuss common problems, such as safety, weapons control and security, and dismantlement. In addition American and Russian scientists discuss laboratory conversion to pursuits more consistent with the absence of either a hot or cold war.

Future Possibilities

I will end by making a few predictions.

First, there is no chance that nuclear weapons will be entirely eliminated in the foreseeable future. Nuclear fission cannot be undiscovered. (Most say "uninvented." I think "undiscovered" is stronger and more accurate.) As long as knowledge of this simple natural process remains in human minds, it will be possible for states to recreate nuclear weapons and, if technology continues to develop as it has in recent years, many of them will be able to do so in a matter of months. It should, however, become possible to find means for capping their numbers in the hundreds rather than the thousands.

Second, nuclear weapons will become delegitimized and marginalized. Most states, including all the large ones, will come to treat nuclear weapons in a fashion parallel to the way they now treat biological and chemical weapons. In particular, they will no longer even consider using nuclear weapons for their own purposes. They will, however, undertake certain preparations to defend themselves against possible use by others, including maintaining a modest nuclear weapons capability purely as a deterrent.

Third, the world's leading nations, including all those now possessing nuclear weapons, will eventually get together and forbid both the use and the threat of use of nuclear weapons by anybody. Hopefully, an effective means for jointly enforcing such a prohibition will be elaborated and organized by the U.N. Security Council.

Fourth, most nations, including all the largest ones, will eventually adopt the view that they have no security problems that cannot be better solved by non-nuclear means and that the salient feature of nuclear weapons is the danger they pose when they are possessed by renegade states. Certain other states may reach different conclusions. Indeed, some relatively small states are located in parts of the world where they arguably do face—or believe they face—problems which currently have no non-nuclear

solution. Such states currently include Israel and Pakistan, and these may be joined by North Korea and Ukraine. Some of these already maintain nuclear forces for handling such problems and others appear to be trying to acquire them. This means that in the future, perhaps even within the next ten years, the critical thinking about nuclear strategy and the crucial decisions about developing and acquiring nuclear weapons will not take place in Washington, Moscow or Beijing, but in such cities as Karachi, Jerusalem, Pyongyang, Bagdad, and, perhaps, Kiev.

Acknowledgments

Most of the essays in this volume were originally printed or presented elsewhere.

MAKING WEAPONS, TALKING PEACE and ARMS LIMITATION STRATEGIES are from *Physics Today*. They are reprinted here by permission.

NATIONAL SECURITY AND THE NUCLEAR-TEST BAN, co-authored with Jerome B. Wiesner, MILITARY TECHNOLOGY AND NATIONAL SECURITY, and THE DEBATE OVER THE HYDROGEN BOMB are reprinted with permission. Copyright 1964,1969,1975 by The Scientific American, Inc. All rights reserved.

THE ARMS RACE AND THE FALLACY OF THE FIRST MOVE, THE ORIGINS OF THE LAWRENCE LIVERMORE LABORATORY, and STRATEGIC RECONNAISSANCE, the last co-authored with G. Allen Greb, are reprinted with permission from *The Bulletin of the Atomic Scientists*. Copyright 1969, 1975, 1977 respectively, by the Education Foundation for Nuclear Science, 6042 South Kimbark, Chicago IL 60637, USA.

NEGOTIATIONS AND THE U. S. BUREAUCRACY was my contribution to a seminar sponsored by the Roosevelt Center for American Policy Studies and published as chapter XII in *A Game for High Stakes*. Ballinger Books, edited by Leon Sloss and M. Scott Davis.

THE COMPREHENSIVE TEST-BAN NEGOTIATIONS is entitled *Making Weapons, Talking Peace*. It was published by Basic Books in 1987 as one of a series of scientific memoirs in a program sponsored by the Alfred P. Sloan Foundation. Reprinted with permission.

NUCLEAR DETERRENCE AND THE MILITARY USES OF SPACE is reprinted with the permission of *Daedalus*, Journal of the American Academy of

Arts and Sciences, from the issue entitled, "Weapons in Space, Vol, I: Concepts and Technologies," Spring 1985, Vol. 114, No. 2.

The Chapter entitled WHY SDI?, was originally entitled STAR WARS. It was co-authored with Sanford Lakoff, and is reprinted with permission of the *Journal of Policy History*, Vol. 1, No.1, 1989, Pennsylvania State University Press.

REMARKS ON MINIMUM DETERRENCE was presented at a conference held in 1990 at the University of California Lawrence Livermore National Laboratory under the auspices of the Department of Energy. The U.S. Government retains a non-exclusive royalty free license to this material.

THE NUCLEAR ARMS RACE: PAST, PRESENT, AND FUTURE was presented at a symposium at the Royal Swedish Academy of Sciences, December 5, 1992, on the occasion of the 50th anniversary of the first nuclear chain reaction. It is reprinted here with permission.

Index

About the Author

I n the worldwide arms race, Herbert F. York has earned the reputation as one of America's leading statesmen. Recruited before his twenty-first birthday to the Manhattan Project, he was eventually appointed founding director of Lawrence Livermore Laboratory, one of the nation's chief weapons labs. Later, he was made Ambassador to the Comprehensive Test-ban Negotiations (1979–81) in Geneva where he labored to control the weapons of mass destruction he helped create.

A member of the von Neumann Committee that mapped the development of intercontinental missiles, York was also first chief scientist of the Advanced Research Projects Agency. In 1958, President Eisenhower appointed him first Director of Defense Research and Engineering.

Widely published, York is the author of the influential *Race to Oblivion* (1970), *The Advisors: Oppenheimer, Teller and the Superbomb* (1976), and a searching memoir, *Making Weapons, Talking Peace* (1989).

Professor York's academic career has been entirely at the University of California, where he earned his doctorate in physics and was then appointed to the Physics Department at Berkeley. In the early sixties, he served as Chancellor of the San Diego campus, where later he was professor of physics and Dean of Graduate Studies. In the late eighties, York was Director of the Program in Science, Technology and Public Affairs. He is now Director Emeritus of the Institute on Global Conflict and Cooperation.

Among the author's well-deserved distinctions and honorary doctorates, York received the 1994 American Physical Society Leo Szilard Award for "his outstanding leadership in efforts to control nuclear weapons." In 1993, York was presented with the Public Service Award by the Federation of American Scientists and earlier he was given the AEC's E.O. Lawrence Memorial Award.